Cross Emergence Of COVID-19

SHOAIB AHMAD

Copyright © 2024 Shoaib Ahmad

All rights reserved.

ISBN: 9798322566915

DEDICATION

This book is dedicated to the Health Care Providers all over the world who faced the SARS-CoV-2, took care of its victims, looked after them and provided support when no one else could.

CONTENTS

	Acknowledgments	i
0	Introduction	1
1	Emergence Functions of Information Manipulating Dynamical Systems	5
2	The Cross-Emergence Function	36
3	Pandemics as Emerging Dynamical Systems	59
4	The Emergence of COVID-19	80
5	Cross Emergence of COVID-19	105
6	The First 100 days and the 166 weeks of COVID-19 in five (5) chosen countries	125
7	Emergence, Cross Emergence, and Indices of Divergence of the Variants of SARS-CoV-2 in the USA	157
8	Self-Organizing, Emerging, Cross Emerging SARS-CoV-2	185
9	List of References	199
10	The Author page	212

ACKNOWLEDGMENTS

1. The World Health Organization (WHO) is gratefully acknowledged for the meticulously collected and transparently shared data on the two episodes of the severe acute respiratory syndrome coronavirus (SARS-CoV-2) during 2002-4 and from 2019-onwards on its web sites. The 2002-4 data was extracted from its daily situation reports; the complete lists of the COVID-19 cases are available on the WHO web page https://data.who.int/dashboards/covid19/.

2. Our World In Data, a project of the Global Change Data Lab, a registered charity in England and Wales, is gratefully acknowledged for the data on the sequencing of the variants in the USA from their web page https://ourworldindata.org/grapher/covid-variants-bar.

INTRODUCTION

The Book aims to investigate the pandemics caused by the severe acute respiratory syndrome croronavirus-2 (SARS-CoV-2) by employing an information-theoretic model that utilizes the probability distribution functions constructed from cumulative numbers of the infected cases in local and global environments as a function of the duration of pandemic.

Graphs of the Logistic growth of the accumulating numbers of the SARS-CoV-2 infected persons, provide the emergence-defining parameter in the form of the inflexion points of the logistic curves that quantify the process of growth of the pandemic, in diverse situations, as a function of its duration. The Shannon entropies calculated from the normalized, cumulative probability distributions provide the second diagnostic tool to unambiguously quantify the emergence of information-generating and manipulating dynamical system of the pandemic.

An **Emergence Function** is defined with the above two parameters- the inflexion point of the logistic graph of the growth of population of the cases and the maximum of the cumulative Shannon entropy. These Emergence Functions are evaluated for each episode of the growth of virus-infected cases in the cities, countries, and the WHO-designated regions of the world. Similarly, the globally accumulated cases generate global Emergence Functions.

The emerging profiles and trends of the pandemic will be described in terms of the Emergence Functions evaluated from the probability distribution functions for each episode, stage, and the wave of SARS-CoV-2 cases.

Another, information-theoretic diagnostic function, **Cross Emergence Function** (CEF) is introduced for comparative evaluation of the pandemic's emergence in diverse environments occurring during the same time scale. The phenomenon of cross emergence is defined by utilizing cross entropies generated by the two probability distributions $p(\zeta)$ and $q(\zeta)$ as a function of time ζ. The cross entropies define and quantify the divergence of the competitive probability distributions. The set of cross entropies treat the distributions $p(\zeta)$ and $q(\zeta)$, alternatively as the true and the predicted distributions. The CEFs are defined and evaluated through the inflexion points derived for each distribution and the set of cross entropies for the two distributions.

Emergence and Cross Emergence Functions provide the bedrock of the information-theoretic model to describe the emergence and evolution of the pandemic, for shorter and longer durations, in smaller geographical locations of cities as well as the global regions of the world.

An **Index of Divergence** (IoD) will be introduced as the absolute value of the difference of the two Cross Emergence Functions derivable from the twin distributions $p(\zeta)$ and $q(\zeta)$ that represent temporally coincident waves/stages of the pandemic. The Index of Divergence will be used to characterize the divergent profiles of the cross-emerging stages and waves of the pandemic in different countries and regions. It will be used as the relative emergence measure in Chapters Three to Six. The IoD will also be employed to serve as a quantitative tool to measure divergence between the variants of SARS-CoV-2. The IoDs will be demonstrated to be an evolution-quantifying measure for the emerging variants of SARS-CoV-2 in Chapter Seven.

Two distinct episodes of SARS-CoV-2, in 2002-4 and for the 166 weeks duration of 2019-2023, are investigated using the above-mentioned information-theoretic tools i.e., Emergence Function, Cross Emergence Function and the Index of Divergence.

The first two Chapters introduce the Information generating and manipulating dynamical systems and define the Emergence and Cross

Emergence Functions, and the Index of Divergence for the Source-Reservoir-Sink model to simulate the pandemics.

Chapter Three analyzes the first episode of SARS-CoV-2 pandemic in 2002-4 with emphasis on four countries, China, Hong Kong, Singapore, and Canada.

Chapters Four and Five present the Emergence and Cross Emergence of COVID-19 in the six(6) WHO-designated Regions of the world.

Chapter Six is dedicated to the Emergence and Cross Emergence of COVID-19 in the five chosen countries, USA, Germany, Spain, Korea, and China, of the three(3) WHO-Regions of the Americas, Europe, and W Pacific. Two timescales are investigated, the First 100 days and the 166 weeks of the pandemic. The IoDs derived for the two different time scales identify the scale invariant character of the pandemic.

Chapter Seven is focused on the investigations of the Emergence of Variants of Concern and the Evolution of SARS-CoV-2 in the USA. USA was chosen for the study of the evolution of the virus as it has the best available data on the sequencing of the variants, transparently shared through multiple websites. USA had the largest number of the cases as compared with other countries, and therefore, the availability of the data on the sequencing of the variants helped in understanding the evolutionary profiles of the variants in a country with ~102 million cases in 166 days out of the global total of ~758 million.

Chapter Eight sums up the self-organizing pandemic's evolutionary trail based on the investigations of the Emergence and Cross Emergence of SARS-CoV-2 in the USA. It is suggested that the pandemic in the USA represents global trends. The exception of the unique and diversifying features of the pandemic in the W Pacific region have been indicated and discussed in Chapters 4 to 6.

COVID-19 emerged, locally and globally, as a self-organizing, information generating, and manipulating dynamical system.

It demonstrated sensitivity to initial conditions that led to the unique emergence profiles.

Pandemic spread within diverse environments of countries and regions with profiles of Emergence → Cross Emergence depicted by the information-theoretic tools.

Information-theoretic diagnostic tools to quantify Evolution are defined as the Emergence and Cross Emergence Functions and the Indices of Divergence.

The emergence profiles of COVID-19 displayed self-similarity and scale invariance, at micro and macro-scales.

COVID-19 started with the emergence of SARS-CoV-2 in its original form. The virus evolved, mutated, and developed strategies of transmission and survival. The evolutionary trails are identified in the book.

On the question of the endemic nature of the virus; the present, ongoing trends of the Omicrons suggest that SARS-CoV-2, as a dynamical system, has self-organized towards an Open-ended evolution, as opposed to the End-directed Emergence.

CHAPTER 1
EMERGENCE FUNCTIONS OF INFORMATION MANIPULATING DYNAMICAL SYSTEMS

1.1. Information-theoretic Source-Reservoir-Sink Model. The structures, formulations, analyses, and the use of mathematical models of pathogen transmission during pandemics have been extensively studied. As early as 1927, the paper by Kermack and McKendrick on the mathematical theory of pandemics studied the emergence of epidemics and pandemics relating the initial few cases of the virus-infected persons with the growing population of the cases over an extended period [1]. Temporal evolution of the cumulative number of cases are studied so that the patterns of growth are understood and can be analyzed. Epidemic and pandemic models are employed for the analyses of virus-driven challenges. New and novel types of infection-related data demand that the models provide the understanding of epidemic processes for improved disease control. Statistical models of epidemics play a crucial role in understanding and predicting the spread of infectious diseases. These models combine mathematical principles with the collected data to estimate parameters and assess the impact of various interventions and help project the mechanisms of infectious disease progress, providing insights into epidemic outcomes and informing the public about the healthcare interventions [2-20]. Another, related mathematical model of the growth of population of living species that has relevance to the emergence of pandemics is to investigate the dynamics of population growth through the mathematical models of predators and prey [21-26]. The case of the growth of human populations in an early paper in 1920, discussed the mathematical representation of population growth in the United States since

1790 [22].

Mathematical model presented by Renshaw [24], can be employed in predicting the emerging population behavior, assessing the impacts of the environment, and learning the species limiting strategies. The Renshaw model provides the resources for understanding the dynamics of biological populations across spatial and temporal dimensions. In his approach, the deterministic and stochastic models can be considered together. The spatial effects involve geographical restrictions on the species mobility that may impact population development. Most aspects of population dynamics are covered, for example, population growth through logistic processes, competition among the growth limiting forces, fluctuating environments, and the spatial systems. The temporal aspects emerge through conservatory practices that affect the velocities of spread. His model treats epidemics as spatial branching structures that may display specific temporal dimensions.

In this book, an information-theoretic model is introduced that treats epidemics and pandemics as entropy generating, sharing, and manipulating dynamical systems. The emergence of such a dynamical system will be evaluated through the normalized probability distributions of the population of the virus-infected persons, referred to as the 'cases' using the World Health Organization (WHO) terminology. The model will present the optimum productive capabilities of the Coronavirus pandemic in terms of the Emergence Function, to be defined later in this chapter. The Emergence Function will be evaluated by utilizing the spatial character of the emerging pandemic and quantified through the cumulative Shannon entropy. The temporal dimension will be introduced through the logistic graphs of the emerging population of the cases. The inflexion points of the population's logistic growth curves will play the essential role in determining the temporal character of the information generating dynamical system of the pandemic in specific locations and at certain times. The amount of information generated through the logistic graphs of the sequences of growth during the various stages in a locality or in different regions, will help to categorize the pandemic and its relative severity. The Emergence Function utilizing the inflexion point and the optimally generated Shannon entropy will be shown to characterize the diversity of the virus infected population growth on local scales that eventually help build the global pandemic's dynamic profile. This chapter will introduce the Emergence Function.

May's seminal work suggested that the first-order nonlinear difference equation arising in the biological, economic, and social sciences, exhibits the dynamical behavior, from stable points to the bifurcating hierarchies of stable cycles and to the onset of chaos [23,25,26]. May demonstrated that the nonlinear difference equation produces logistic maps of the population value at any time step related to its value at the next step depending on the rates of growth. The population level at a given time can be treated as a function of the growth rate parameter and the previous time step's population level. For too slow a growth rate, the population may die out i.e., negative growth leading to the extinction of the species. Higher growth rates might settle the system toward stable values or fluctuate across a series of population explosions. In the context of this book, the extension of the regular growth to bifurcations and chaos will not be included. The logistic equation can lead to the respective maximum, limiting capacities of population of the pandemic, with the environment-specific rates of growth. The graphs of the logistic growth of population provide the essential analytic tools that can identify the patterns of growth of the Coronavirus infected cases in various places, countries, and regions. The logistic equation will be used in this chapter and the following ones to evaluate the emergence functions of the infected population along with the associated cumulative Shannon entropy.

The two episodes of the Coronavirus pandemics started in 2002 and 2019, with a few cases, spread to larger communities, and travelled across the globe through all possible means of available transportation and communication utilized by the infected persons. In the case of the 2019 pandemic, the constantly mutating and adapting to the different environments, the virus self-organized and its mutations spread all over the globe at the rates allowed by the ambient conditions.

In addition to the logistically growing virus infected populations, the information-theoretic model assumes that the pandemics emerge as self-organizing dynamical systems [27-38]. The emergence of these dynamical systems can be described by adapting the information-theoretic tools developed by Claude Shannon [39,40]. Shannon's model of information generating and transmitting systems consisting of the Transmitters, Channels, and Receivers, can be extended to describe the characteristic emergence of the probability distributions generating constituents and components of the dynamical systems. The emergence of these interactive dynamical systems is dependent on adapting to the conditions and restrictions imposed from within and from outside the system. However, the analysis of such systems, by utilizing the information-theoretic techniques, can be done if and only if the restrictions imposed by Shannon on the delivery of the 'original message' are removed [40]. For example, instead of coding and preserving the 'original message' emanating from the Transmitter, the 'modulations of the original message' should be allowed by the dynamical system. This can happen if the Channel is allowed to modify and manipulate it before transmission to the Receiver. The systems with the desired information manipulative capabilities will lead to the information-theoretic basis for understanding the patterns and profiles of the emergent structures and their interlinked properties. The information-manipulating dynamical system requires that the Source/Transmitter of these systems generate the 'original message,' to be modified and manipulated by the Channel/Reservoir at the succeeding stages. The Source delivers the 'original message' to the Reservoir that may modulate and modify it before transmission to the Receiver. This general introduction forms the basis of the information-theoretic model of the emerging dynamical system (DSs) that generate, modify, and manipulate the 'original message' in the Source-Reservoir-Sink (SRS) model [41]. The SRS model provides the tools and techniques that control the modified signals being transmitted in the *Source → Reservoir → Sink*, shown in Figure 1.1. The model is valid for the conservative flows as well as the growth/generative cases where the population may grow (as in pandemics) or shrink.

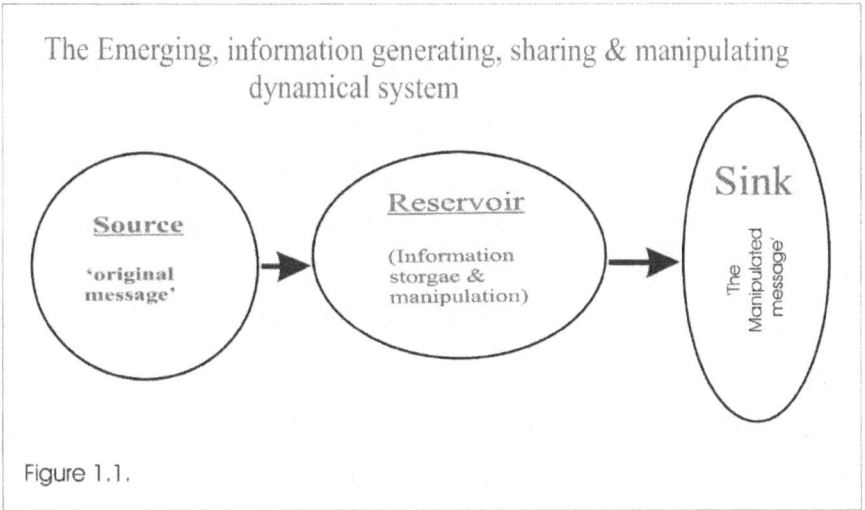

Figure 1.1.

Figure 1.1. An information generating, sharing, and manipulating dynamical system composed of the 3-components, the Source, Reservoir, and the Sink (SRS) model, is shown that can produce the manipulated outcomes, rather than transmitting the 'original message'. The Sink receives 'The manipulated message' after passage through the twin information manipulating channels of the Source-Reservoir and the Reservoir-Sink.

The present chapter deals with the emerging characteristics of dynamical systems consisting of the 1-, 2- or 3-components. Logistic growth models of such systems have been proposed that lead to the emergence of complex systems [25]. Starting from the formulation of the model from a single box with a dripping faucet [42], the level of complexity and the number of components of the systems gradually increased. The objective is to develop tools to characterize the diversity of the pandemic. The information-theoretic properties will be shown to be capable of forming the building blocks of the systems with the increasing number of components that can generate optimally productive diversity. After introducing the essential dynamical parameters, the Emergence Functions will be evaluated from the appropriate probability distributions functions that yield the distinctive cumulative Shannon entropies for each of the information generating, modulating, and receiving constituent of Source-Reservoir-Sink [41].

The broad range of the emergence of dynamic phenomena in physical systems and nature have been investigated as dynamic cooperative phenomena [43,44], as non-equilibrium dissipative structures [45,46], and as the emerging science at the edge of order and chaos [47]. The emergence of a pandemic can also be treated as the emergence of a complex system [47]. In such cases, the emergence of the coherent structures is based on the self-organizational processes that can be identified through the generation of data of the infected cases whose patterns have the fractal property of self-similarity and sensitivity to initial conditions [48-56]. Starting from Chapter Three of this book, the information-theoretic descriptions of the emergent cumulative probability distribution functions of the pandemics in different countries and the regions will be shown to project the self-similarities and sensitivity to initial conditions. These fractal characteristics will also be represented in the cumulative entropic and the cross entropic profiles at all population levels of the Coronavirus pandemic in the local, regional or the global stages.

In this chapter, the emergence profiles of the SRS dynamical system will be evaluated by utilizing the cumulative probability distributions, generated by each of the inter-connected components, which can identify the distinguishing features of the information generating, sharing, and modifying of the 'original message' at each distinctive stage of transmission through the dynamical system. The set of Shannon entropies, calculated from the probability distribution functions, present the framework for the information-theoretic modeling of the diverse systems that emerge and evolve as a function of their specific conditions and constraints. The Emergence Function will function as a measure of the extent of the temporal existence of entropy (information) generating capabilities of each of the constituent (Source-Reservoir-Sink) of the model. Emergence Functions will characterize and quantify the relative-emergent behavior of the constituents of the dynamical systems evaluated from the inflexion points of the respective logistic curves of the increasing pandemic populations and the associated cumulative Shannon entropies.

A range of emerging dynamical systems can be described by the application of the general features and functions of the SRS model. The general model can be extended to include those dynamical structural arrangements that allow the initial or the 'original message' to be manipulated and modified according to the constraints imposed on the system. The self-organizing mechanisms adapted by the Coronavirus pandemics during 2002-4 and for the 166 weeks of 2019-2023, are the prime examples of the general framework of the information generating and manipulating dynamical systems across the diverse physical landscape [57].

1.2. The Emergence Function λ of Dynamical Systems. The cumulative Shannon entropies generated by the constituents and components of information manipulative dynamical systems define and determine the relative performance of the interactive constituents to display their emergence. The entropic description of such a DS can be defined by employing the cumulative probability distribution functions that effectively describes the emerging characteristics. It leads to the evaluation of the Emergence Function that is defined through the maximum entropy and the inflexion point of the associated logistic graph. The Emergence Function λ will be used in the later chapters as the measure of the amount and the pace of the pandemic's information generative capabilities.

The set of cumulative entropies is derived from the normalized cumulative probability distributions for each component of the Source-Reservoir-Sink. Each of the evaluated probability distribution depends on the rates of the in-flow and out-flow of information. Let the probability of a component of the SRS system be defined as a function of the distribution steps or stages ζ be referred to as $p(\zeta)$. The associated uncertainty of occurrence of a certain event at step ζ is $\ln(1/p(\zeta))$. This function has special characteristics and has also been named 'surprise.' It has been referred to as the Kolmogorov measure of uncertainty [58]. The special characteristics associated with $\ln(1/p(\zeta))$ are related with the fact that its value logarithmically increases for the lower values of $p(\zeta)$. The constantly increasing $\ln(1/p(\zeta))$ springs the 'surprise' for the successively decreasing probabilities [59]. It is zero when the outcomes are certain i.e., 0 or 1. Shannon cumulative entropy is evaluated as the sum of the product $\{p(\zeta)\ln(1/p(\zeta))\}$ known as instantaneous entropy for each incremental step of the measure ζ. The sum over all instantaneous entropies is the well-known Shannon entropy or Information [40]

$$H = -\sum_{\zeta} p(\zeta)\ln(p(\zeta)) \equiv \sum_{\zeta} p(\zeta)\ln(1/p(\zeta)) \qquad (1.1).$$

It must be pointed out that $\ln(1/p(\zeta))$ and H are dependent upon the measure ζ and consequently on the rate of the generation of $p(\zeta)$. The diverse configurations of the dynamic system will yield the varying values of $\ln(1/p(\zeta))$ and H corresponding to the rates of information exchange and the increasing or decreasing number of the information generation steps/stages ζ.

The Emergence function λ evaluated from H will be employed to describe the DSs and to enquire whether the stages and sequences of the emergence profiles of the emerging DSs lead to the End-directed or the Open-ended Emergence? [57]. The direction of Emergence will be the consistent theme of the book.

As the first step, the Emergence Function λ of the 1-Box configuration with a dripping faucet will be defined using the entropic profile followed by the investigations of the 2- and 3-Box configurations of information-generating dynamical systems.

1.3. A Box with a dripping faucet. The probability paradigm, used in this chapter for the study of dynamical systems, relies on the generation of random as well as the controlled distributions. Let the box with a dripping faucet [42,57] as shown in Figure 1.2, generate the probability distribution functions that are plotted for the box filled with a sharable/transferable material at random and regular rates. In Figure 1.2(a), the material, in the box was removed by letting the valve to randomly drain between 0 and 1/2 of the quantity remaining in the box at each successive step ζ. Four sequences of the removal of the material that change the Box's state {1} to {0}, are shown. The box in the inset with {1} implies the initial, filled state. The transition of the state {1} to the final state of the empty box {0} constitutes the Box's emergence as a DS with the dripping faucet generating the randomly generated probability distribution functions $p(\zeta)$ in 1.2(a) and at the regular rate of ½ in 1.2(b).

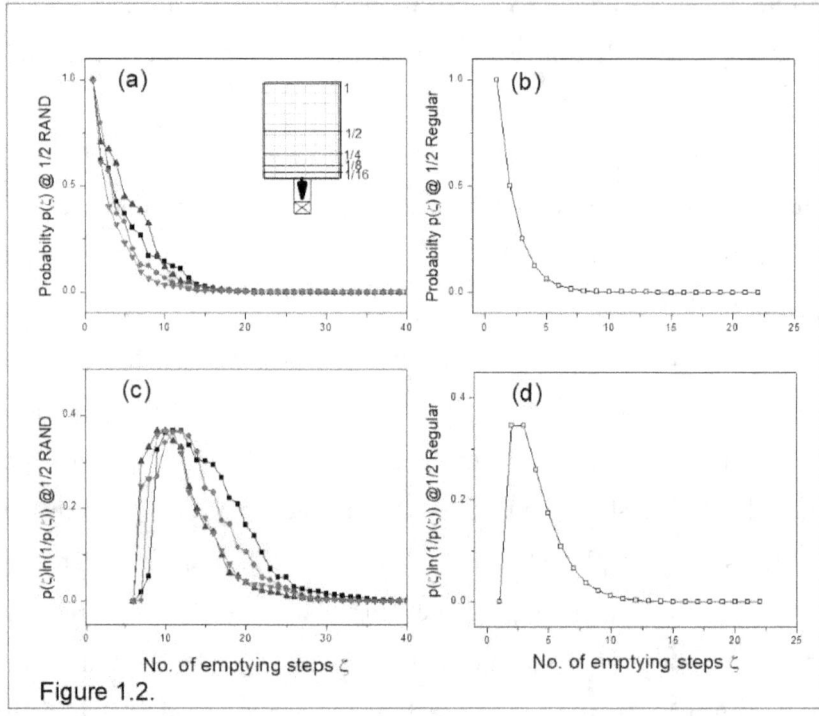

Figure 1.2.

Figure 1.2. The emptying profile of the Box with a dripping faucet. (a) Four randomly generated graphs of the probability distribution function $p(\zeta)$ are plotted for each emptying stage ζ. A random number generator produces a number between 0 and 1/2. At each successive step, the Box is emptied by the quantity determined by this random number generator. (b) The box emptied at a regular, fixed rate $\alpha = 1/2$ shows the emerging probability distribution $p(\zeta)$. (c) The instantaneous entropy $p(\zeta)\ln(1/p(\zeta))$ shown for the four random graphs in (a). (d) The $p(\zeta)\ln(1/p(\zeta))$ for the graph with regular rate shown in (b). **Note:** The area under the graphs in (c) and (d) yields the respective cumulative Shannon entropy $\sum p(\zeta)\ln(1/p(\zeta))$.

In Figure 1.2(a) and (b) the probability distributions $p(\zeta)$ were generated as a function of the box-emptying ζ-steps. In 1.2(c) and (d), the graphs for the instantaneous entropy $p(\zeta)\ln(1/p(\zeta))$ are constructed as a function of ζ by using the probability distributions $p(\zeta)$ for the random and the regular rates α, shown in Figures 1.2(a) and (b).

The above example demonstrates a one-component DS, generating random and regular sequences for emptying the Box with a dripping faucet.

Such a simple DS can produce complex profiles including the transition to chaotic regimes as May and other researchers have pointed out [25,27-38]. The information-theoretic paradigm presented above, built around the probabilistic behavior of a single Box, identifies the tools for the study of the dynamics of the inter-dependent Emergence of a variety of dynamical systems.

Figure 1.3(a) shows the sum of the successive instantaneous entropies from Figure 1.2(c) and (d), to yield the cumulative entropy $H = \sum p(\zeta)\ln(1/p(\zeta))$ as a function of the information generating stages ζ. The cumulative entropy graphs with random rate α, approach the approximately steady maximum cumulative entropies beyond the limit $\zeta \sim 15$. The regular emptying rate α generates comparatively lower entropy that maximizes around $\zeta \sim 5$. The approach to maximum entropy characterizes the emergent functional dependence of the dynamical systems to be discussed later in the chapter.

The emergence profiles of the cumulative entropies H of the regular and the random probability distribution functions graphed in Figure 1.3(a) show similar growth trends. Initiated by the nonlinear growth mechanisms culminating towards the maximized cumulative entropic capacities. The entropic graphs (H versus ζ) can also be described by the nonlinear Logistic equation used for the population growth. This aspect will be helpful in evaluation of the Emergence Function to be introduced later in the chapter.

The emergence profiles of the cumulative entropies H of the regular and the random probability distribution functions graphed in Figure 1.3(a), show similar growth trends. Initiated by the nonlinear growth mechanisms culminating towards the maximized cumulative entropic capacities. The graphs of the cumulative entropies are the emerging profiles of the information generating dynamical system that provide the growth of entropy as a function of the number of the emptying steps ζ. We treat these graphs of emergence, as the entropic equivalent of the logistic graph with the holding capacity as maximum entropy H_{max} and the inflexion point ζ_0 defined along the ζ-axis at the half (½) of the holding capacity H_{max}. The Emergence Function is defined as

$$\lambda = \zeta_0 * H_m \tag{1.2}$$

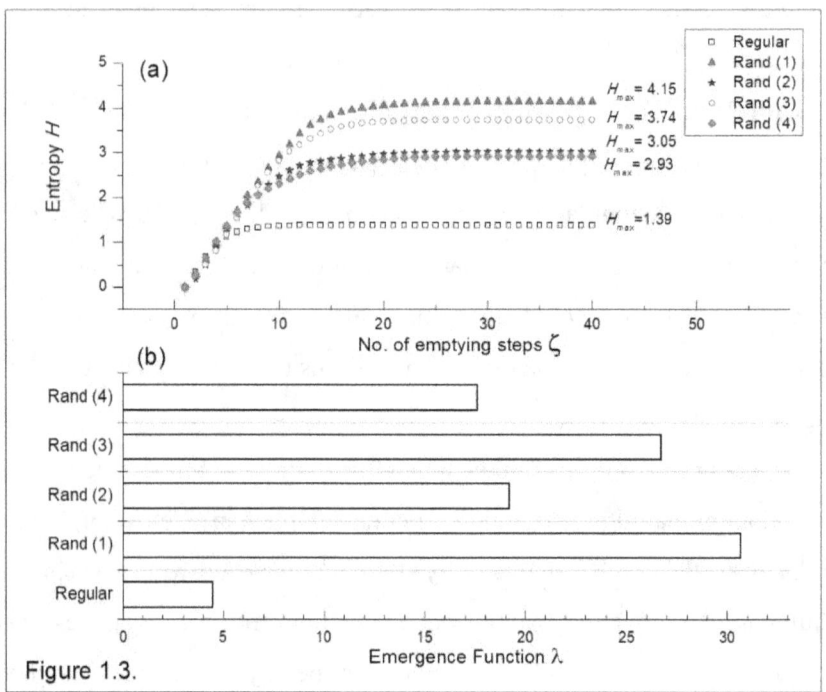

Figure 1.3.

Figure 1.3. Cumulative Entropy of the Box with the dripping faucet. (a) The cumulative entropy $H = \sum p(\zeta)\ln(1/p(\zeta))$ for the regular and the four random flow probability distributions is plotted as a function of ζ. The maximum generated entropy H_{max} for each graph is shown. (b) The Emergence Function λ for each simulation of the Dripping Box is plotted as bar graphs using Eq. (1.2).

In Figure 1.3(b), the ζ_0s are obtained by fitting the Boltzmann functional method for the sigmoidal graphs, on the data of the entropic graphs in 1.3(a). Alternatively, the inflexion point ζ_0 can also be obtained from the logistic curves of the emerging populations. The former method is preferred, as will be shown later.

In Figure 1.3(b), the five bar graphs, one for the regular, and four for the random emptying of the Box, are shown. The regular flow with the rate $\alpha=1/2$ has the minimum $\lambda=4.5$. The random flows generate $\lambda \sim 17.6$ to 31 with the average of four flows $<\lambda> \sim 23.6$. Whereas the regular flow ensures the approximate completion of the emptying within ~7-9 steps, the random flows equivalent require ~20 steps. The other noticeable feature relates to the inflexion point ζ_0, the regular flow has the minimum $\zeta_0 \sim 1.4$. The average $<\zeta_0> \sim 6.7$ for the random rates of flow. The approximate, average divergence between the regular and random rates of flow is ~1:4 demonstrated through the ratio of the Emergence Functions λ(regular) and λ(random).

The dynamic, information-theoretic model of the 1-Box with a dripping faucet will be extended to the 2-and 3-Boxes. The multi-component DSs may operate under the varying conditions and constraints to yield specific objectives and goals. These objectives are described in the following sections.

1.4. The 2-Boxes model of a Source with a Sink. The addition of a collector box to function as the Sink to the Source with a dripping faucet discussed above, can yield a dynamical system displaying uniquely divergent, Emergence characteristics as compared with the 1-Box-Source. Two probability distributions are now generated; one for the Source sharing its contents and the other for the recipient Sink, by utilizing the methods and the tools developed in the previous sections.

The set of the two emerging probability distributions of the twin-component DS (Source-Sink), are constrained by the condition $p(\zeta) + q(\zeta) = 1$. The two probability distribution functions are shown in Figure 1.4(a). The Source with the dripping faucet is shown to generate two different sequences of the outward flow at the regular rate $\alpha = (1/2)^m$, where $m=2$ and 5. The distribution $p(\zeta)$ can be represented as the controlled output from the Source for each of the next step $(\zeta + 1)$ as

$$p(\zeta + 1) = \alpha p(\zeta) \qquad (1.3).$$

The Sink's filling profile emerges as the growth curve with the rate γ that can be evaluated from the Logistic equation for the population growth. This nonlinear equation is used to obtain the emergent Sink's probability distribution $q(\zeta)$ as a function of ζ. It represents the dependence of the sequential growth of population where each growth step is related to the previous value, through the rate of growth γ. The rate γ can be obtained from the fitting of the data in Figure 1.4(b) with the iterations of the Logistic equation. The equation generates the sequential set of the population growth of Sink $q(\zeta + 1)$ at the step $(\zeta + 1)$ being related to the preceding $q(\zeta)$ at ζ [25]

$$q(\zeta + 1) = \gamma q(\zeta)(1 - q(\zeta)) \qquad (1.4).$$

The emptying of the Source can be represented by the set $\{1..., 0\}$ while the Sink's consequent set for the filling distribution is $\{0..., 1\}$. The model ensures at each dynamical stage ζ, the conservation requirement implies $p(\zeta) + q(\zeta) = 1$. In Figure 1.3(a) and (b), the emptying of the Source is shown for the two controlled rates $\alpha = 1/4$ and $1/32$. The starting probabilities at $\zeta = 0$ for the Source and the Sink are 1 and 0. The Source undergoes a transition $\{1\} \rightarrow \{0\}$, subsequently the Sink follows $\{0\} \rightarrow \{1\}$. Emptying of the Source to the Sink leads to the transformation of the composite, 2-Boxes dynamical system as $\{1,0\} \rightarrow \{0,1\}$ for $\zeta \rightarrow \infty$.

In the Figure 1.5, the cumulative entropy $H = \sum p(\zeta) \ln(1/p(\zeta))$ is plotted as a function of the increasing number of the information transfer steps ζ for the set of the Source and the Sink of Figure 1.4 for the slower rate $\alpha = (1/2)^5$.

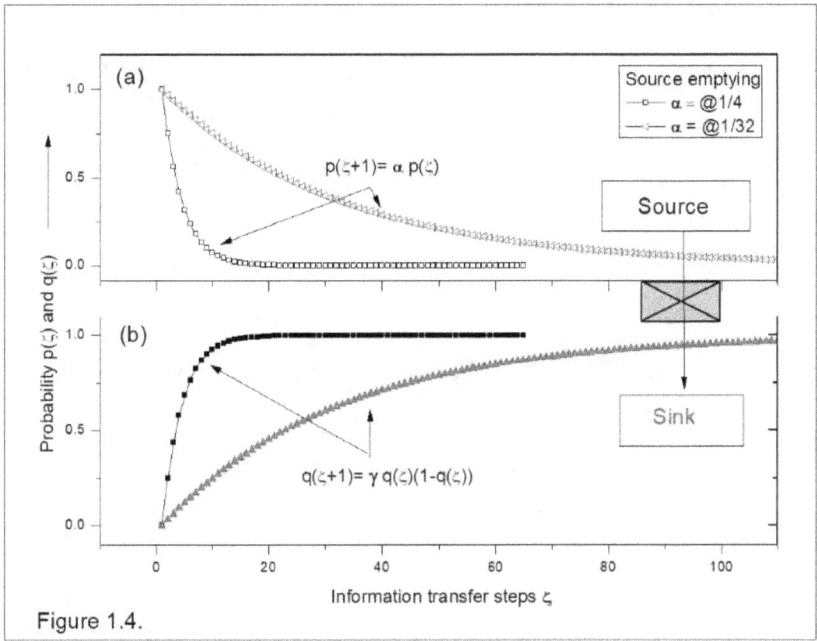

Figure 1.4.

Figure 1.4. The 2-Boxes model of the Source and Sink. (a) The mirror image probability distributions $p(\zeta)$ and $q(\zeta)$ are plotted as a function of the information transfer steps ζ. Two graphs for each distribution are plotted, one with the rate of flow $\alpha = @1/4$ and the other with $\alpha = @1/32$. (b) The emptying of the Source is done via Eq. (1.3) while Eq. (1.4) can describe the filling profile of Sink as $q(\zeta + 1) = \gamma q(\zeta)(1 - q(\zeta))$.

Three noticeable features from the two graphs of the cumulative entropy for the Source and the Sink, in Figure 1.5 are:

- The Sink generates higher entropy than the Source up to a critical value of ζ labelled in the figure as ζ_{EEL}. The two entropies equalize at this limit designated as the entropy equalizing limit-EEL. The significance of this limit is that it represents the equalizing as well as the local maximizing of the two interdependent entropies. It is a unique stage in the evolution of the entropic profile of any information generating and sharing DS composed of two components. For the thermal systems, the entropy generation stops at ζ_{EEL} signaling the approach to the state of equilibrium.

- For the non-thermal systems, the second feature relates to the reversal of the twin entropic profile for $\zeta > \zeta_{EEL}$. The Source starts to generate consistently higher entropy than the Sink. This is the second noticeable feature of the transition of the Source from the state {1} towards the empty state {0}.
- Figure 1.5 demonstrates that for the entire dynamic range of the information transfer stages from $\zeta = 1$ to $\zeta \gg \zeta_{EEl}$, the Source-generated entropy is always greater than the respective cumulative entropy of the Sink for the entire range where the system transition {1,0} to (0,1} occurs; $H(Source) > H(Sink)$ for $\zeta \to \infty$.

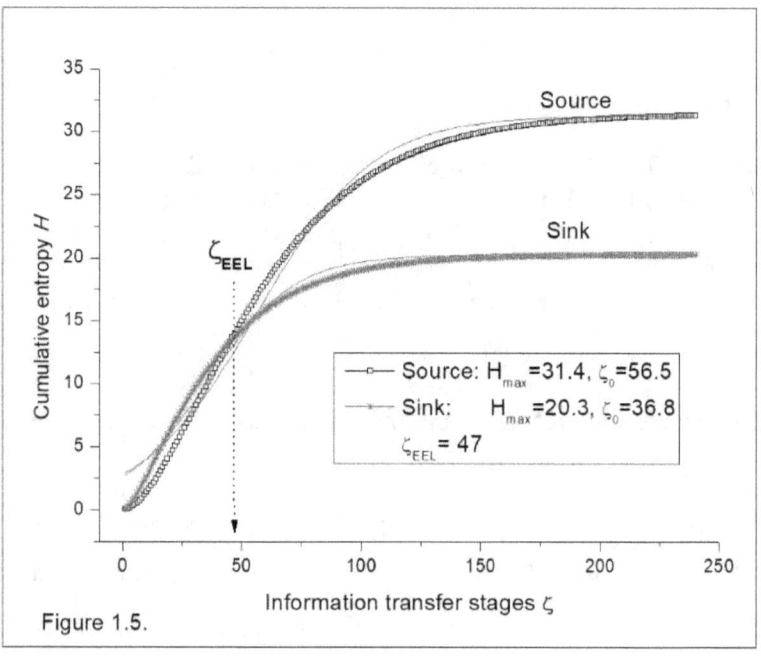

Figure 1.5.

Figure 1.5. The entropy equalizing limit ζ_{EEL}. Two graphs for the cumulative entropy $H = \sum p(\zeta)\ln(1/p(\zeta))$ of the Source-Sink system are shown as a function of the information transfer steps ζ. The entropy equalizing limit ζ_{EEL} is indicated, with its value shown in the boxed inset. The graphs are fitted with the sigmoidal curves to derive the inflexion points ζ_0. The Emergence Functions λ for the Source is 1774, and 747 for the Sink.

The twin entropies of the Source and the Sink in 1.5 approaching their respective approximate maxima for the values of $\zeta \to \infty$. These cumulative entropic growth curves can also be described by employing the Logistic equation (1.4), with the respective rates of growth being determined by curve fitting.

1.5. Logistic growth of the emerging populations. The emerging populations of the Sink of the 2- and the 3- or higher number of Boxes, can be described through the Logistic equation [25]. It relates the population at each stage $(\zeta + 1)$ with its value at the previous step ζ. This nonlinear equation of $q(\zeta)$ as a function of ζ for the population growth, is used to evaluate the rate of growth of population. Similarly, the cumulative entropy graphs as a function of the measure of the growth ζ, can also be described as a logistic graph with the inflexion points ζ_0 determined through curve fitting, as will be shown below.

For the present discussion, Eq. (1.4) has been presented as a difference equation. It can also be defined as a differential equation in the form $dq(\zeta)/d\zeta = q(\zeta)[\gamma - kq(\zeta)]$. The ratio of the two variables of the equation yield $\gamma/k = L$, where L is the holding capacity or the maximum achievable population for the dynamical system. The equation can be solved for the initial population q_0 as $q(\zeta) = q_0 L/\{q_0 + (L - q_0)\exp(-\gamma\zeta)\}$, to fit the population growth data [27,28,36]. For the normalized probability distributions, the holding capacity $L = 1$, the equation takes the form for the initial value of population $q(\zeta = 0) \equiv q_0$

$q(\zeta) = 1/\{1 + A \exp(-\gamma\zeta)\}$; where $A = \{(1 - q_0)/q_0\}$ (1.5).

Equation (1.5) generates the set of the values of $q(\zeta)$ as a function of ζ for the Sink's cumulative population from q_0 to $\sum_{\zeta=0}^{max} q(\zeta)$. Fitting the data with the iterative $q(\zeta)$ from Eq. (1.5) can determine the rate of population growth γ. Logistic equation (1.5) describes a dynamical system that is unstable at $\zeta = 0$ and moves towards a stable solution. Such a stable solution exists for the holding capacity $L=$ maximum allowed capacity of the population as $q(\zeta) \to 1$. The typical Logistic graph has three significant states: (1) The unstable state at $\zeta = 0$ with $q(\zeta = 0) \equiv q_0$, (2) Another unstable state at the inflexion or the mid-point $q(\zeta_0) = 1/2$ and (3) the stable state for $q(\zeta_{max}) \to 1$.

The two unstable states (1) and (2) and the stable state (3) of the population growth curve of $q(\zeta)$ versus ζ, determine the shape and the lateral extent of the Logistically defined population-growth curve of the Sink. The condition (2) introduces an important relationship at $\zeta\{q_{(\zeta)=1/2}\} \equiv \zeta_0$ in the form of $q(\zeta) = 1/\{1 + A\exp(-\gamma\zeta)\} = 1/2$, this lead to the rate of growth of population in terms of A and ζ_0 as $\gamma = \ln(A)/\zeta_0$. Therefore, by evaluating ζ_0 of the growth curves, the corresponding rates of growth γ can be calculated.

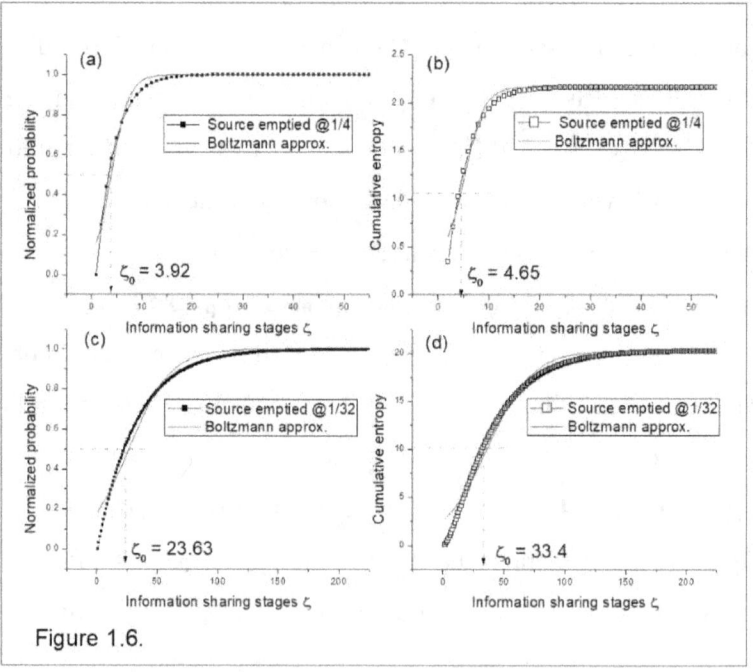

Figure 1.6.

Figure 1.6. Logistic growth of estimation of the Sinks in the 2-Boxes configuration. (a) & (b). The data from Figure 1.3(b) is used in Eq. (1.5) that yields the rates of growth γ shown in the boxed insets of (a) and (b) for the Source emptying @1/4 and 1/32, respectively. The respective ζ_0 are shown in the figures. (c) & (d) have the entropic graphs plotted as a function of ζ for the corresponding cumulative population values in (a) and (b).

The inflexion point ζ_0 provides the temporal dimensional parameter that uniquely characterizes a dynamical system's population growth, or the cumulative entropic growth profiles. This function will be used extensively for evaluating the Emergence Function λ of the evolving, emerging dynamical systems. Along with the entropy H_m which defines the approach to the stable state, the inflexion point ζ_0 provides a relative measure of emergence for different constituents of the system.

In Figure 1.6, the Source-Sink configuration of section 1.4 is evaluated from the perspective of the filling profile of the Sink for the two rates of the Source emptying @1/4 and 1/32. Figure 1.6(a) and (b) are the normalized cumulative population densities from Figure 1.4(b). The curve fitting of the cumulative population of $q(\zeta)$ using Eq. (1.6), the ζ_0 values are determined for 1.6(a) and(b). In Figure 1.6(c) and (d), the cumulative entropies are plotted for the population growth shown in 1.6(a) and (b). The ratio of {ζ_0(cumulative entropy) / ζ_0(cumulative probability)} is ~ 1.18 for the rate $\alpha=1/4$ and 1.41 for $\alpha=1/32$. The above figure provides an alternative route to evaluate the inflexion points from the cumulative entropic graphs. Such an exercise has relevance when the initial number of the infected cases may not be uniquely determined, as will be the case during COVID-19, in the later chapters.

1.6. The 3-Boxes Source-Reservoir-Sink (SRS) model [41]. The 3-Boxes model can now be introduced where a third box is inserted between the Source and the Sink of the previous section, to function as an information storing and manipulating Reservoir. It is called Reservoir due to its intermediate position and the inherent material retention and transmission capabilities. The flow of material and the associated information in the 3-Boxes-SRS can be assigned in the similar, one-way direction as in 2-Boxes' case; from the Source → Reservoir → Sink. Let the initial state of the 3-Boxes be {1,0,0}. It indicates that the Source is the originator or the initiator of the information, as in the 1-Box and the 2-Boxes configurations. The Reservoir starts from an empty state {0} and return to the empty state {0} at the end of the transfer of all material and the related information. The Sink, as in the earlier 2-Boxes configuration, is depicted as the ultimate recipient of all the transferred material and generates its specific cumulative entropic profile for the states from {0} to {1}. The final state of the Source-Reservoir-Sink system will change from {1,0,0} to {0,0,1}.

The set of the three probability distribution functions for the Source-Reservoir-Sink combination is plotted in Figure 1.7(a). The synchronous, controllable valves are made to operate at the chosen rate of emptying $\alpha \equiv \alpha_1 = \alpha_2 = 1/2$ at each information generating step ζ. Consequently, the three probability distribution functions $p(\zeta), r(\zeta)$ and $q(\zeta)$ are generated as shown in Figure 1.7(a). The distributions are constrained by the conservation rule, as was applied to the 2-Boxes. For the 3-Boxes, at every stage of emergence the conservation rule is strictly obeyed i.e., $p(\zeta) + r(\zeta) + q(\zeta) = 1$.

In Figure 1.7(b), three graphs for the instantaneous Entropy $p(\zeta)\ln(1/p(\zeta)), r(\zeta)\ln(1/r(\zeta))$, and $q(\zeta)\ln(1/q(\zeta))$ are plotted for each incremental step of information generation and transfer, for the whole sequence of the dynamical events. The state of the system transforms from the initial state of the SRS system {1,0,0} to {0,0,1}.

Figure 1.7(c) plots the cumulative entropies $H_{Source} = \Sigma p(\zeta)\ln(1/p(\zeta))$, $H_{REservoir} = \Sigma r(\zeta)\ln(1/r(\zeta))$, and $H_{Sink} = \Sigma q(\zeta)\ln(1/q(\zeta))$ for the system operating at the same rates of flow $\alpha_1 = \alpha_2 = 1/2$. However, the DS generates different probability distributions with the associated entropies. These cumulative entropies maximize for $\zeta \sim 10$. This aspect will be further elaborated in the next section where the equality of the rates $\alpha_1 = \alpha_2$ does not hold.

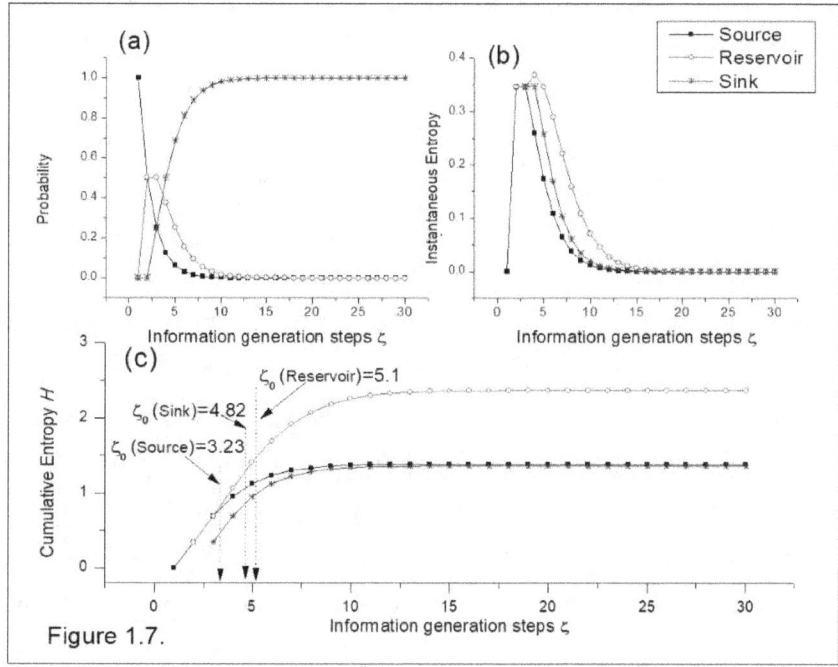

Figure 1.7.

Figure 1.7. Source, Reservoir and Sink (SRS) model. (a) The probability distribution functions for the 3-component SRS are shown. The emptying of the Source and Reservoir occur at the rate $\alpha_1=\alpha_2=1/2$. (b) Instantaneous entropy is plotted as a function of the number of information generation steps. (c) The cumulative entropy H is shown for the graphs in (b) as a function of ζ. The inflexion points ζ_0 of Source, Reservoir and Sink are indicated with arrows.

The emerging features of the DS consisting of a Source, a Reservoir and a Sink are graphically displayed in Figure 1.7. The nature and the significant aspects of this mode of transmission of information from Source-to-Reservoir-to-Sink can be summed up as:

- Information can be transmitted at any rate α_1 from Source-to-Reservoir and at the rate α_2 from Reservoir-to-Sink. Figure 1.7(a) is constructed with the two equal rates of transfer of material from the Source and the Reservoir with $\alpha_1 = \alpha_2 =$

1/2. However, these can be different as will be demonstrated in the next sections. The difference amongst the two rates can create markedly different entropy-generating regimes.

- It is shown that the 'original message' from the Source is manipulated by the Reservoir during transmission to the Sink. The three cumulative entropies in 1.7(c) are indicative of the manipulative character of the Reservoir.
- Independent of the rate of transmission by the Reservoir, the transmitted signal to the Sink and the associated entropic profile will always be different from the 'original message' transmitted by the Source. This happens irrespective of the fact that the two valves may or may not be operating at the same rate $\alpha_1 = \alpha_2 = 1/2$.
- The Sink is the recipient of the manipulated signal from the Reservoir. As far as the Sink is concerned, the Reservoir appears as the source of its Information. The variations in the two, distinct rates of information exchange, α_1 for the Source→Reservoir, and α_2 for the Reservoir→Sink, can produce either a Free-Flow regime as shown in Figure 1.7, the Source-dominated or the Reservoir-dominated regimes that will be discussed in the next sections.
- The Sink is 'blind folded' vis-à-vis the Source-to-Reservoir communications. On the other hand, the Sink is the recipient of the net material flow and the associated information which is dependent on the two rates of flow α_1 and α_2. These together determine the basic characteristics of an interconnected, information-sharing 3-Box configuration.
- The requirement of the transition $\{1,0,0\} \rightarrow \{0,01\}$ ensures that only the one-way communication is discussed with the direction of the information flow Source→Reservoir→Sink. However, the feedback loops may exist in nature or even in man-made information sharing devices and may need to be considered in evaluation of such systems' responses.
- Other possibilities can also be considered for the transitions and material transfer/retention in the 3-Boxes configuration. The above, rather simplified version of the transfer of the state of the system from $\{1,0,0\}$ to $\{0,0,1\}$ helps us construct the model in its simplest form. The other options may exist whenever an ongoing transition stops, or a state of pause emerges. For example, one could have $\{0.6, 0.25, 0.15\}$ as the instantaneous state of the system that implies the

transitions {1...., 0.6} for the Source, {0......0.25} for the Reservoir and {0......0.15} for the Sink. An incomplete transition can also exist in nature and may also be dealt with by using the above developed SRS model.

- In the case of pandemics treated as emerging dynamical systems, the condition of the conservation $\{p(\zeta) + r(\zeta) + q(\zeta) = 1\}$ is not valid. The probability distributions as a function of the information transfer steps ζ will follow the inequality $p(\zeta) \neq r(\zeta) \neq q(\zeta)$, as a rule. However, the general framework of evaluating the DS remains the same, as will be highlighted in Chapters 3-to-7.

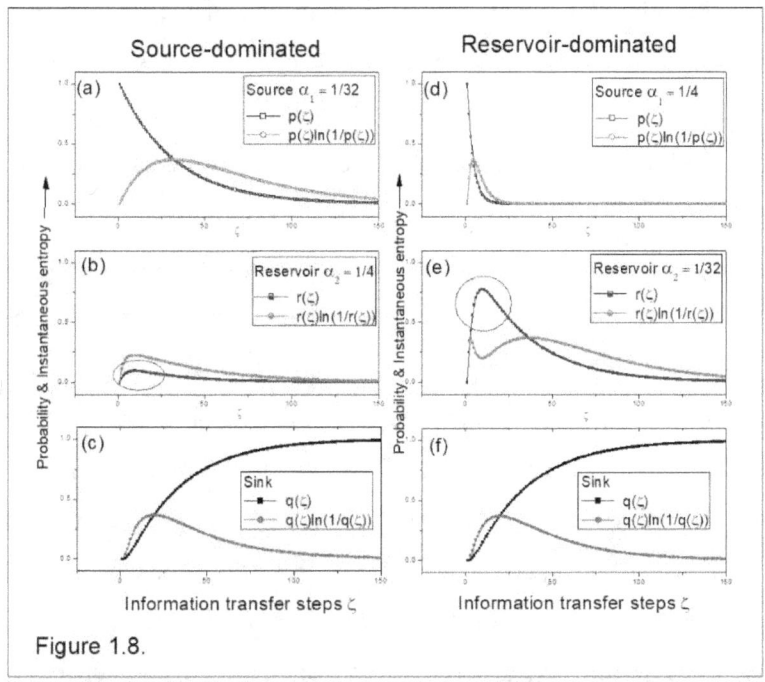

Figure 1.8.

Figure 1.8. Source and Reservoir dominated SRS. The figure shows the three probability distribution functions $p(\zeta)$, $r(\zeta)$ and $q(\zeta)$ and the corresponding instantaneous entropies of the Source, Reservoir and Sink plotted for two different rates as a function of the number of information transfer steps ζ. In (a) the Source-to-Reservoir rate $\alpha_1=1/32$. In (b) $\alpha_1=1/4$. In (c) and (d) the corresponding reversed values for the for the Reservoir-to-Sink rates are rate $\alpha_2=1/4$ and $\alpha_2=1/32$. Figures (e) and (f) show the evolving Sink's probability distributions and the instantaneous entropy as a function of ζ.

1.7. Manipulation of the rates α_1, α_2 of the SRS. Figure 1.8 has the profiles of the probability distribution functions $p(\zeta)$, $r(\zeta)$, and $q(\zeta)$ and the corresponding instantaneous entropies $p(\zeta)\ln(1/p(\zeta))$, $r(\zeta)\ln(1/r(\zeta))$, and $q(\zeta)\ln(1/q(\zeta))$ of the Source, Reservoir and the Sink plotted for two SRS configurations with different rates of information flow $\alpha_1 \neq \alpha_2$. Figure 1.8(a), (b) and (c) are for the combination of the two rates $\alpha_1 = 1/32$ and $\alpha_2 = 1/4$. Here, a slowly emptying Source with a fast-draining Reservoir lead to the Source-dominated dynamical system of the SRS. The second configuration shown as Figure 1.8(d), (e) and (f) has a fast-draining Source→Reservoir with $\alpha_1 = 1/4$ and a slower rate $\alpha_2 = 1/32$ for the Reservoir→Sink. This combination of the two rates generates the Reservoir-dominated system.

Unequal rates of flow produce two scenarios; the first one is where the Source empties slowly and hence it generates higher information. Such a situation may favor certain physical systems that are designed to retain material or information for longer durations. On the other hand, if the build-up of the retention capacity of the Reservoir is desired, that may require relatively slower rates of transmission to the Sink. In addition to the manipulation of the signal received by the Reservoir, the emergence of the specific, desired properties of the Source-Reservoir-Sink configuration may endow the End-directed Emergence characteristics to the self-organizing ensembles of carbon cages [57]. These are highlighted in Figure 1.8(a)-(f) where the relative rates of information generation by the Source and Reservoir are varied in the inverse order.

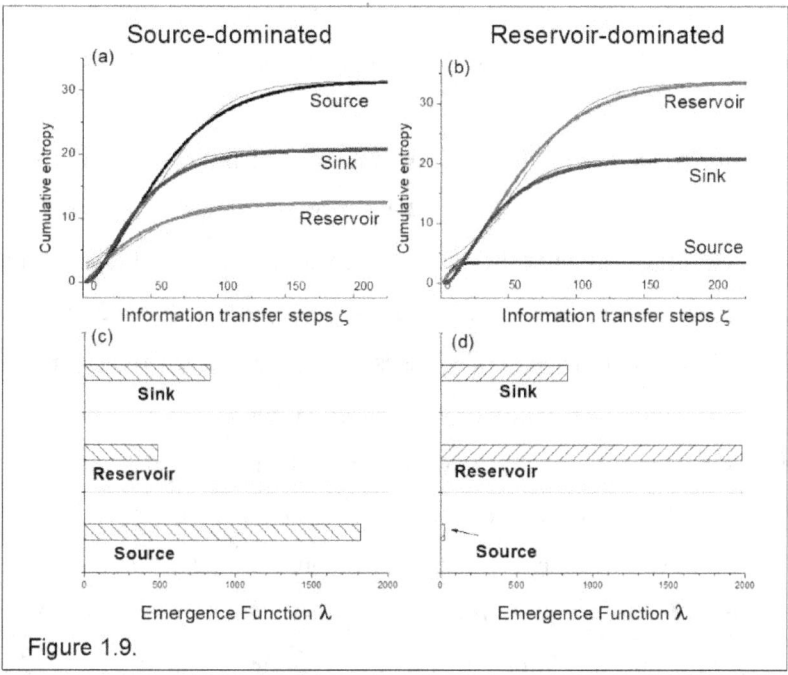

Figure 1.9. (a) The cumulative entropies $\sum p(\zeta)ln(1/p(\zeta))$, $\sum r(\zeta)ln(1/r(\zeta))$, and $\sum q(\zeta)ln(1/q(\zeta))$ for the Source dominated SRS(1) configuration are shown. The data from Figure 1.8 for the instantaneous entropies were used. (b) The three entropy graphs for the Reservoir-dominated SRS(2) configuration show where the Reservoir generates higher entropy than the Source. (c) The Emergence Function λ demonstrates the dominating Source. (d) The set of Emergence Function λ of the Reservoir-dominated SRS system clearly shows that the Reservoir effectively dominates over the Source. The Reservoir increases the extent of information manipulation to generates higher Information for the Sink as opposed to that from the Source.

Figure 1.9 utilizes the data for the instantaneous entropies of Figure 1.8 and plots the respective cumulative entropies $\sum p(\zeta)ln(1/p(\zeta))$, $\sum r(\zeta)ln(1/r(\zeta))$ and $\sum q(\zeta)ln(1/q(\zeta))$ for the two configurations i.e., the Source-dominated and the Reservoir-dominated in 1.9(a) and (b). The Emergence Functions λ are plotted as bar graphs for the Source-dominated system in 1.9(c). The Emergence Function λ demonstrates the dominating Source. The ratio of λ(Source): λ(Reservoir)=4:1. In Figure 1.9(d), the set of the three Emergence Function λs of the Reservoir-dominated SRS system clearly shows that the Reservoir completely dominates over the Source. with λ(Source):λ(Reservoir)=1:80. The Reservoir increases the extent of information manipulation to generate much higher Information for the Sink as opposed to that from the Source. On the other hand, the Sink emerges as the information-receiver with approximately similar ratio with the dominant player of the SRS system. The ratios are, λ(Source):λ(Sink)~2.2:1 for the Source dominated case and λ(Reservoir):λ(Sink)~2.4:1. Figure 1.9 demonstrates the two performance measuring indicators.

The ratio of the entropies for the SRS(1) is 31.47:12.52:20.74 and 3.45:33.64:20.74 for the SRS(2). It can be noticed that the Sink's cumulative entropy remains the same for the two SRS configurations where the rates of information exchange were reversed. The Reservoir generates 10 times larger entropy than the Source in the SRS(2) configuration as compared with the 2.5 times $H(Source)/H(Reservoir)$ for the SRS(1).

The corresponding inflexion points ζ_0 for the two sets of SRSs show that the dominant constituent of both the SRS(1) and SRS(2) generate approximately similar $\zeta_0 \sim 58$ as Source(1) and Reservoir(2).

The Reservoir-dominated SRS limits the role of the Source with much smaller $\zeta_0 \sim 7.23$ and $H_m \sim 3.5$.

This leads to the significance of the intermediate stage of the Reservoir in information manipulation. The Source generates the 'original message,' and the Sink receives it in the modified form after passing through the Reservoir. In the Shannon model, the Channel is considered as the source of noise, degradation, and alteration of the 'original message.' In the SRS model, presented here, Channel or the Reservoir plays the similar but 'desirable,' signal-manipulation role for the emerging dynamical systems. Simultaneously, it acts as the repository of information, the material to be transferred, or the associated information. Some, or all its features can be controlled for 'desirable' effects. In the case of pandemics, these aspects have been investigated and highlighted here, in this chapter. and will be discussed further in the next one.

1.8. Summary. To determine and define the emerging features of the change inducing elements or processes, dynamical systems have been characterized in the preceding sections, by the information-generating and manipulating setups. The emergence defining Emergence Function λ is derived from the system's response to the continually stimulating initiator in terms of the emerging probability distribution functions. These distributions depict the impact of the initiator on the entire system and its constituting components. Shannon entropy or the Information can be calculated for each component and for the entire system as a function of the dynamic measure of the information sharing stages ζ. This dynamic measure determines the duration of the evolving structures and their temporal dimension. Every entropy generating component of the DS can be tracked and analyzed through the cumulative entropic profile H as a function of ζ, during and at the final stages of emergence.

The Emergence Functions λ are defined for the evolving dynamical systems and their components to serve as the quantitative measure of system's Emergence. The cumulative entropies of each constituent of the SRS provide the first comparative assessment. Their maximum entropic values H_m yield the relative performance indicator. Secondly, the effective number of the stages is identified with the inflexion point ζ_0 of the emerging entropic profile. Together these are used to evaluate the Emerging Function $\lambda = \zeta_0 * H_m$. This function combines the entropic and temporal dimensional parameters of emergence.

- In the case of a single Box with the dripping faucet, the profiles of emptying at regular and random rates generate Emergence Functions to serve as the dynamic measure of the amount of and the pace of entropy generation.

- The 2-Boxes configuration of the Source with a Sink combination provides insight into the mechanisms of information generation and sharing. The emptying of the Source and the filling of the Sink generated two distinct, related probability distribution functions $p(\zeta)$ and $q(\zeta)$. The growth curve of the Sink represented by $q(\zeta)$ is the mirror image of the Source's emptying probability distribution function $p(\zeta)$. The distribution functions are set to obey the conservation paradigm $p(\zeta) + q(\zeta) = 1$ throughout the dynamical exchange of the information that converted the initial state of the system $\{1,0\}$ into $\{0,1\}$.

- The 3-Boxes, Source-Reservoir-Sink (SRS) was introduced as the network of information-generating, manipulating, and sharing components. The Source-Reservoir-Sink display a self-organizational character. The SRS configuration endows the Reservoir the ability of storage, manipulation, enhancement, or degradation of the information (or material) received from the Source. Probability distributions are constructed that are used to calculate Shannon entropy. The 3-component Source-Reservoir-Sink classifies and quantify various aspects of the evolving, change-inducing conditions, and the constraints that define the emerging features and the final outcomes.

- The Reservoir can have memory, fixed or variable. In the SRS model, only the Reservoir may interact and exchange information (or material) with the Sink. The Sink is the recipient of the material and information. The interactive participation of the Reservoir with Source and Sink is the core of the SRS model. The Sink may receive (a) material output in various shapes, configurations, and forms, (b) signals that are representative of the processes that occurred in the Reservoir, and (c) useable or waste energy, etc.

- In the case of an emerging pandemic, the growth of the system's population does not follow the conservation rule set out above. Its growth and emergence, however, can be determined through the Emergence Function λ.

- Coronavirus being a molecular system, can be represented as the form of information that the virus carrier (acting as Source) can transmit to another person. The emerging population of recipients may function as the Reservoir as well as the Source. The groups of infected person becoming the emerging Reservoirs, leading to the epidemic as the Sources and Reservoirs start to grow.

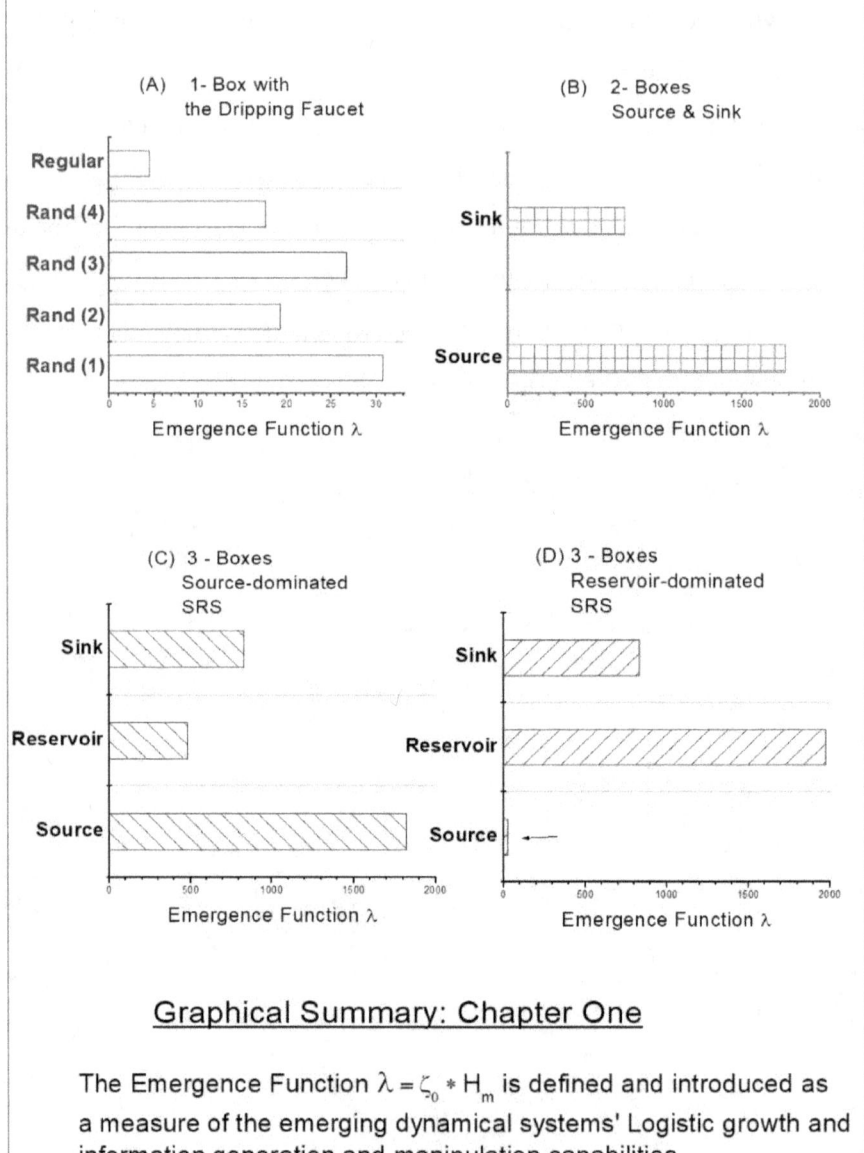

Graphical Summary: Chapter One

The Emergence Function $\lambda = \zeta_0 * H_m$ is defined and introduced as a measure of the emerging dynamical systems' Logistic growth and information generation and manipulation capabilities.

CHAPTER 2

THE CROSS-EMERGENCE FUNCTION

2.1. Introduction. The Emergence Function λ was introduced in the last chapter for the emerging, entropy-generating dynamical systems (DSs). The DSs with single- or multiple-component were shown to generate distinctive probability distribution functions for each constituent, which generated the respective cumulative entropies as a function of the information-generating and sharing steps ζ. The Emergence Function was defined as a measure of the spatial and temporal emergence for each constituent of the dynamical system, under the ambient conditions. The variations in the dynamic parameters of the system, for example the varying rates of flow of information among the constituents, which induce changes in their respective Emergence Functions (λs), were elaborated in Chapter One. The essential features of the Emergence Function are described here, once again, due to the emerging requirement of another Cross Emergence Function to provide the intra-constituent comparisons. Let us re-establish the parameters of the Emergence Function λ.

1. A solitary box filled with a liquid and fitted with a dripping faucet (section 1.3) can generate a well-defined probability distribution for the regular flow rate, from the initial filled state {1} to the empty state {0}, or it may generate an array of distributions for the random rates of flow. The emergence functional profiles were plotted in Figure 1.3.

2. The 1-Box DS can be subjected to the condition of material conservation if a collector is added to the system. It was shown that the 2-Boxes analogy (section 1.4.) of the first Box with a dripping faucet acting as

the Source, and a second box as filling bucket of the Sink, generates the probability distributions $p(\zeta)$ and $q(\zeta)$ that fulfill the condition $p(\zeta) + q(\zeta) = 1$ at each transition step ζ for the information transfer. This condition is equally valid for the regular as well as the random flows. Similarly, the conservation principle can be extended to the 3-Boxes equivalent, the Source-Reservoir-Sink (SRS) where $p(\zeta) + r(\zeta) + q(\zeta) = 1$, as described in detail in sections 1.6 and 1.7 of Chapter One.

3. The above-mentioned condition of conservation is violated when the SRS model is extended to the Coronavirus pandemic. The ensembles of the initial few cases of the Coronavirus-infected persons function as the Source during the initial stages of the locally emerging epidemics or the globally spreading pandemics. Multiplication and the spreading of the virus to a larger population generate the Reservoirs with the increasing number of cases. If the probability distributions of the increasing number of the infected cases in the Sources and Reservoirs can be constructed as $p(\zeta), r(\zeta)$, then the conservation rule of Chapter One do not apply, and neither do the respective cumulative sums of the cases are equal. Conservation rule of the number of the infected cases does not hold in pandemics and one deals with the growth of population of the cases. The patterns and profiles of the growth might vary in various places, countries, and regions, and produce the location-specific Logistic curves with the inflexion points ζ_0 for the growing ensembles of the virus-infected cases. Their respective emergence profiles, however, can still be represented by Eq. (1.2) for the Emergence Function $\lambda = \zeta_0 * H_m$.

The above-mentioned population-growth model of the diversity of the patterns of the increasing number of infected persons in different places under varying conditions during pandemics, demonstrate the need for another, divergence-quantifying function. Such a function is needed for comparative analyses of the emergence of diverse ensembles of virus-infected cases in separate places during the same temporal space. The cross emergence function can be formulated by utilizing the cross entropic divergences generated for the specific periods of time, by utilizing the probability distributions of the emerging populations, under different conditions [60-69].

A function that serves the above-mentioned purpose, can be defined as the Cross Emergence Function (CEF) between the pairs of the probability distribution generating constituents of the DSs. It employs the sets of the twin cross entropies of the interacting constituents along with their constituent-specific, growth-defining parameter-the inflexion point ζ_0.

2.2. Cross entropies and the Cross Emergence Function.

The information-theoretic technique of developing the model for the Emergence Function λ can be adapted and extended to comparing the individual profiles of each of the population growth stages and the modes of Coronavirus pandemic in diverse regions and countries. The extended model is based on evaluating cross entropies of the chosen sets of the twin emerging populations. The Cross Emergence Function will be derived by using the cross entropic profiles for the analyses of the pandemic in separate places, under different conditions and environment, during the same temporal regimes. Such a function will help in identifying the evolutionary trails of the Coronavirus in the later chapters.

The Cross Emergence Functional model, developed here, will utilize the set of the cross entropies $H(p|q)$ and $H(q|p)$ as opposed to the cumulative entropies H_m that were used for defining the Emergence Function λ in Eq. (1.2). The set of the two inflexion points $\zeta_0(p)$ and $\zeta_0(q)$ will be obtained from the respective cumulative entropic graphs using their Logistic growth profiles for the distributions of the populations in the case of pandemics. Alternatively, as in the case of the interconnected Boxes of the dynamical systems, the emerging cumulative entropic graphs for the distributions of the information sharing Boxes, can also be used for evaluating the respective

ζ_0s.

The method applies equally well to the single stage or the successively increasing number of stages of the pandemic. For each stage of pandemic, the cumulative probability distribution functions $p(\zeta)$ and $q(\zeta)$ will be constructed from the number of the emerging cases for any two, comparable, chosen places or environments. The growth of the number of cases during a certain period is represented by the cumulative distribution function of the accumulating cases. This procedure for constructing the distributions $p(\zeta)$ and $q(\zeta)$ for the specific time or the stage of data collection, plays a key role when multiple regions generate the data of the cases of infected persons. Each region will generate its own cumulative entropy that will lead to the calculations of the associated Emergence Function λ during the specific stage or the period of the pandemic. The comparative intra-regional Emergence Functional (λs) profiles of the pandemic will be augmented by the cross entropic evaluations of the respective distributions conducted during the same periods of time. Cross entropy will be shown to provide an additional diagnostic tool to quantitatively ascertain the entropic divergence between the information generating segments/constituents of a dynamical system.

For the probability distributions $p(\zeta)$ and $q(\zeta)$ of any two emergent populations of the pandemic, which occur within the same temporal space, their cross-entropies can be calculated by assuming one as the 'true' and the other as the 'expected' distribution [60-63]. Cross entropy is built up from the sum of the Kullback-Leibler divergence and the nominal cumulative entropy. The set of the two KL-divergence denoted as $D(p \parallel q)$ and $D(q \parallel p)$ provides an important diagnostic tool, as a measure of the relative entropic divergence for the pairs of probability distribution functions $p(\zeta)$ and $q(\zeta)$, constructed from the data obtained in different conditions during the same period. The KL-divergence is formulated as [69],

$$D(p \parallel q) = \sum_{\zeta} p(\zeta) \ln((p(\zeta)/q(\zeta))) \qquad (2.1).$$

$D(p \parallel q)$ as measure of divergence between the two probability distributions $p(\zeta)$ and $q(\zeta)$, while $D(q \parallel p)$ is similarly defined as $D(q \parallel p) = \sum_{\zeta} q(\zeta) \ln((q(\zeta)/p(\zeta)))$. The set of these twin KL-divergences are employed as the essential component of the corresponding set of Cross-entropies $H(p|q)$ and $H(q|p)$.

In the context of our analyses of the emerging pandemic, cross entropies will be utilized as the comparative, quantitative, and analytical tool for understanding their information-manipulative role played in the growth of the cases by the diverse circumstantial environments of the comparable regions.

Whereas, in information theory, cross-entropic minimization as a Monte Carlo technique for the importance for optimization, approximates the optimal sampling estimator by repeating two phases of the construction or obtaining the two probability distributions and then minimizing the cross-entropy between the primary distribution and a second expected distribution. Here, it is used to learn and illustrate the divergent patterns and profiles generated by the pandemic's emergent cases in various places and regions. The Cross Emergence Function will be constructed based on cross entropies. It will serve as a comparative tool for understanding the evolutionary trajectory of the pandemic of the cases.

The twin Cross entropies define and quantify the divergence of the probability distributions of the two competitive distributions $p(\zeta)$ and $q(\zeta)$ are formulated as [60-69],

Nominal cumulative entropy + KL divergence → Cross entropy

$$H(p|q) = \sum_{\zeta} p(\zeta) \ln(1/q(\zeta)) \equiv H(p(\zeta)) + D(p(\zeta) \parallel q(\zeta)),$$
$$H(q|p) = \sum_{\zeta} q(\zeta) \ln(1/p(\zeta)) \equiv H(q(\zeta)) + D(q(\zeta) \parallel p(\zeta)) \qquad (2.2).$$

The cross entropies $H(p|q)$ and $H(q|p)$ treat the distributions $p(\zeta)$ and

$q(\zeta)$, alternatively as the true and the predicted distributions.

The Cross Emergence Functions (CEF) defined below, as $\mathcal{E}(p|q)$ and $\mathcal{E}(q|p)$ are evaluated through the inflexion points and the cross entropies for the two distributions. The inflexion points $\zeta_0(p(\zeta))$ for $p(\zeta)$ and $\zeta_0(q(\zeta))$ for the distribution $q(\zeta)$ can be obtained by two equally valid methods:

(1) by constructing the Logistic graphs for the population growth of the cases, or

(2) by logistically evaluating the mid-points of the respective cumulative entropic graphs generated by the probability distributions of the cumulative growth of the population.

Both methods can be employed, if these are used consistently, without alternating between the two methods. The two methods to evaluate ζ_0s provide consistent, reliable estimates of the CEF defined in this book as

$\mathcal{E}(p|q) = \zeta_0(p(\zeta)) * H(p|q)$, and

$\mathcal{E}(q|p) = \zeta_0(q(\zeta)) * H(q|p)$ \hfill (2.3).

Cross Emergence Function, as defined above, will be employed, along with the Emergence Function λ to investigate the emerging, information-generating dynamical systems of the pandemics which include the first episode of the Coronavirus in 2002-4 in Chapter 3, and COVID-19 in Chapters 4, 5 and 6. These four chapters are dedicated to the study of the emergence of the two episodes of the Coronavirus pandemics by employing the information-theoretic diagnostic tools of the Emergence Function λ and the pair of the Cross Emergence Functions $\mathcal{E}(p|q)$ and $\mathcal{E}(q|p)$.

Another, comparative Index of Divergence (IoD) can be defined as the absolute value of the difference of the two Cross Emergence Functions for the set of the twin distributions $p(\zeta)$ and $q(\zeta)$, as

$\Delta\mathcal{E} = |\mathcal{E}(p|q) - \mathcal{E}(q|p)|$ \hfill (2.4).

The Index of Divergence $\Delta\mathcal{E}$ will be used to characterize the divergent profiles of the cross-emerging stages of the pandemic in different countries and regions. It will be used as a quantitative tool and the measure of divergence between two competitive emerging variants. It will be one of the resources to identify the routes of the evolution of multiple variants of the SARS-CoV-2 in Chapter Seven. Here, in this chapter, $\Delta\mathcal{E}$ will be employed in the context of the multi-Box model of the emerging dynamical systems, to serve as the comparative, quantitative cross emergence indicator.

2.3. The Emergence Function λ of the SRS with random rates of flow.

The 3-component SRS model of Figure 1.7 in Chapter One is re-plotted in Figure 2.1(a) to (c), that generated the three sets of five (5) simulations of the random flow regimes for the emerging probability distributions. The corresponding cumulative entropies of the Source, Reservoir and the Sink are plotted for the flow of material from Source-to-Reservoir-to-Sink in 2.1(d) to (f). The average of the cumulative entropies is shown with bold symbols along with the others, in Figure 2.1(d) to (f). The emergence profile of the SRS follows the conditions:

- At the start, the Source is full, the Reservoir and Sink are empty. The initial state of the system is {1,0,0}. The process of sharing the material will ensure that the final state of the SRS will be {0,0,1}.
- At each material sharing stage or the step ζ, a random number between 0 and 1 is generated and multiplied with the one quarter of the material in the Source, The resulting quantity is sent to the Reservoir.
- The Reservoir also shares up to ¼ of its contents, similarly randomly chosen, with the Sink. The process continues until the transition from {1,0,0} to {0,0,1} is completed by the dynamical, information sharing system of the SRS.
- All simulations shown in Figure 2.1 are constrained by the conservation rule that was specified in Chapter One. When applied to the 2-Boxes, the rule implies $p(\zeta) + q(\zeta) = 1$; its form for the 3-

Boxes is $p(\zeta) + r(\zeta) + q(\zeta) = 1$.

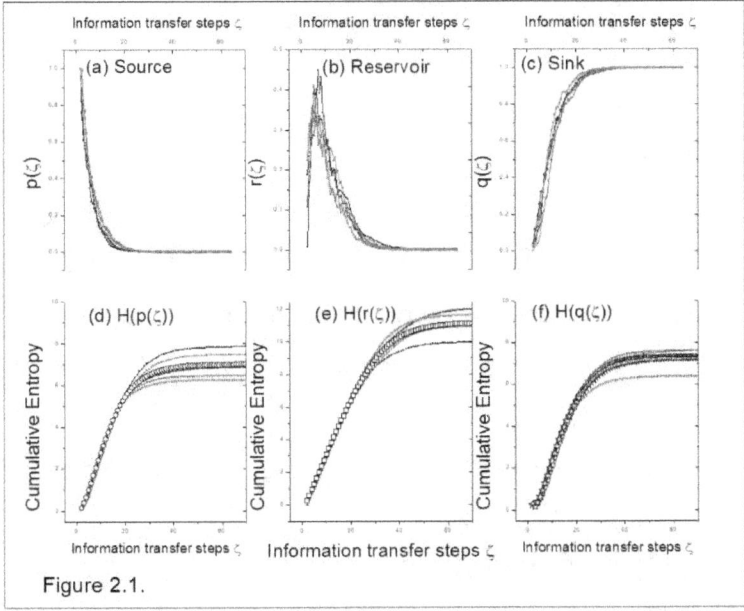

Figure 2.1.

Figure 2.1. The sets of five (5) simulations are plotted in (a) to (c) for the emerging probability distribution functions $p(\zeta), r(\zeta)$ and $q(\zeta)$ as a function of ζ. Starting from the initial state {1,0,0} of the SRS, the dynamical system is set into motion by sending randomly chosen ¼ of the Source contents to the Reservoir. Using the same formula, the Reservoir shares its contents with the Sink, during the transition of the SRS state from {1,0,0} to {0,0,1}. (d) to (f) The emerging cumulative entropic profiles $\sum p(\zeta)\ln(1/p(\zeta))$, $\sum r(\zeta)\ln(1/r(\zeta))$ and $\sum q(\zeta)\ln(1/q(\zeta))$ are plotted for the Source, Reservoir, and the Sink against the information transfer steps ζ.

In Figure 2.1(a), the dynamic profile of the emptying of the Source, represented as the set {1......,0} shows that ~1/2 of the Source had emptied within $\zeta = 0$ to 5 ± 1 steps and ~90% emptied within $\zeta \sim 10$ steps. The corresponding cumulative entropy $H(p(\zeta))$ graph in 2.1(d) identifies the inflexion point $\zeta_0 \sim 13.4$, where the maximum cumulative entropy H_m is halved. The corresponding Emergence Function λ(Source)~93.8, is tabulated in Table 2.1 along with λ(Reservoir) and λ(Sink).

The Reservoir acting as the intermediary between the Source and the

Sink, produces the profile of its set of distributions $r(\zeta)$ shown in Figure 2.1(b). The distributions go through the peaks around $\zeta \sim 10 \pm 4$. These distributions extend well beyond $\zeta > 40$. As opposed to the Source, the Reservoir is a relatively slowing diminishing constituent of the system that generates higher entropy as compared with the other two information-sharing components, the Source, and the Sink. Its Emergence Function λ(Reservoir)~209.6. The Sink, with $\zeta_0 \sim 15.2$ and $H_m = 7.3$, generates λ(Sink)~110. The emerging pattern for the Emergence Functions λs of the SRS with the randomly sharing Source and Reservor is λ(Source) < λ(Sink) < λ(Reservoir).

The Reservoir generates higher cumulative entropy in relatively longer distribution stages.

TABLE 2.1. FREE FLOW REGIMES, EMERGENCE FUNCTION λ

	Source	Reservoir	Sink
ζ_0	13.4	18.7	15.2
H_m	7	11.2	7.3
λ	93.8	209.6	111

A simulation for the same configuration of the SRS with the increased random flow rates of 1/2, instead of the ¼, displays a slightly different pattern of the Emergence Functions that follows λ(Source)~ λ(Sink)< λ(Reservoir). The Source and the Sink λs are of approximately similar magnitudes. For the rates of flow increased to ½ in place of the ¼ used in the simulations presented above, the Emergence Functions get smaller by a factor of 5 as compared with those in Table 2.1, however, the Reservoir always generates higher entropies and correspondingly, larger Emergence Functions. These are due to the conditions of flow that help build the

Reservoir to manipulate the inputs and the outputs. This aspect has already been discussed in the previous chapter in Figure 1.7 that demonstrated this effect. The role of the Reservoir will be further analyzed here in sections 2.5 and 2.6.

2.4. The Cross Emergence Functions $\mathcal{E}(p|q)$ and $\mathcal{E}(q|p)$. The three sets of the six cross entropies $H(p|r), H(r|p), H(r|q), H(q|r), H(p|q)$ and $H(q|p)$ are evaluated and plotted in Figure 2.2 for the three probability distributions, $p(\zeta)$ for the Source, $r(\zeta)$ for the Reservoir and $q(\zeta)$ for the Sink presented in Figure 2.1(a) to (c). The conditions and the constraints of the SRS for Figure 2.1 are maintained for Figure 2.2.

In Figure 2.2(a), the Source and Reservoir generate the set of two cross entropies $H(p|r) = \sum p(\zeta)\ln(1/r(\zeta))$ and $H(r|p) = \sum r(\zeta)\ln(1/p(\zeta))$ as a function of the information transfer steps ζ. For the initial 10 steps i.e., $\zeta \sim 10$, the two cross entropies are approximately equal $H(p|r) \sim H(r|p)$. Beyond $\zeta > 10$, the Reservoir generates larger cross entropy as compared to the Source. For $\zeta \geq 40$, the two cross entropies stabilize around constant values, with a constant ratio of cross entropies $H(r|p)/H(p|r) \sim 0.7$.

Figure 2.2(b) presents the cross entropic profile generated by the respective probability distributions for the information transfer from the Reservoir to the Sink. A factor of ~10 higher cross entropies are generated, as compared with those in Figure 2.2(a) for the Source-Reservoir. The Reservoir-Sink combination generates higher cross entropy than the Sink-Reservoir combination $H(r|q) > H(q|r)$. The inequality continues, from the start towards the larger values of ζ with the cross entropic difference staying almost constant $[H(r|q) - H(q|r)] \sim 24 \pm 2$. The two cross entropies continue to increase and do not stabilize around the maxima like those in Figure 2.2(a). The cross entropic profiles for Reservoir-Sink continue to increase in 2.2(b) as a function of ζ.

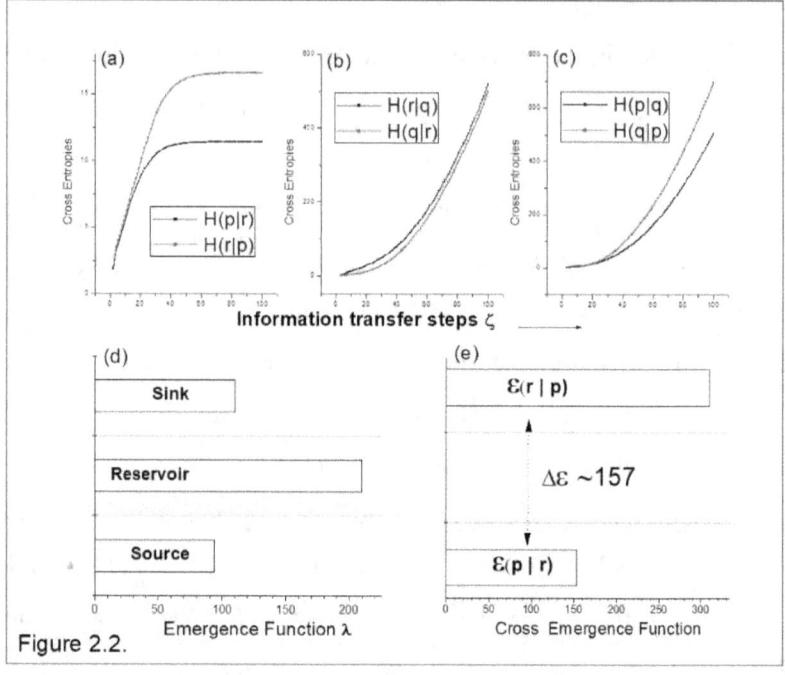

Figure 2.2. Cross entropies are plotted for six combinations of the three probability distributions for the Source $p(\zeta)$, the Reservoir $r(\zeta)$ and the Sink $q(\zeta)$. These are the averages of the graphs shown in Figure 2.1. (a) The Source-Reservoir combination generate the set of two cross entropies $H(p|r)$ and $H(r|p)$ as a function of the information transfer steps ζ. Up to $\zeta \sim 10$, the $H(p|r) \sim H(r|p)$. For $\zeta > 10$, the Reservoir generates larger cross entropy as compared to the Source. (b) The Reservoir continues to generate higher cross entropy as opposed to the Sink, from the beginning and the divergence continues. (c) After the first 20 steps, the Sink-Source combination of distributions produce larger cross entropy and diverge considerable from the cross entropy generated by the Source-Sink combination. (d) The set of three Emergence Functions λs are plotted as bar graphs: with λ(Source)< λ(Sink)< λ(Reservoir). (e) The Cross Emergence Functions $\mathcal{E}(p|r)$ and $\mathcal{E}(r|p)$ are plotted for the Source to Reservoir combination only. Index of Divergence for the Reservoir - Source combination is shown $\Delta\mathcal{E} = |\mathcal{E}(r|p) - \mathcal{E}(p|r)| \sim 157$.

Figure 2.2(c) demonstrates that during the initial 20 steps, the Source-Sink combination of the cross entropies are approximately equal $\{H(q|p) - H(p|q)\} \sim 0$. This cross entropic profile $H(q|p) \sim H(p|q)$ can be interpreted by noting that within $\zeta=0$ to ~20, the Source had emptied ~95% and the Sink had almost filled as shown in Figure 2.1(a) and (c). Beyond the $\zeta>20$ stage, the Sink-Source combination produces the continually divergent cross entropies. Beyond this stage, the Source gets farther away in terms of the cross entropic distance.

Figure 2.2(d) has the Emergence Functions λs plotted as bar graphs. The Reservoir generates the largest λ. The inequality of the Emergence Functions follows the pattern $\lambda(\text{Source}) < \lambda(\text{Sink}) < \lambda(\text{Reservoir})$.

The cross entropies $H(p|r)$ and $H(r|p)$ for the Source-Reservoir combination demonstrate the approach to maxima for $\zeta>20$ where the mutual cross entropic difference $\Delta H = [H(r|p) - H(p|r)]$~constant. That indicates the system approaching the stable states of cross emergence. In Figure 2.2(e), the bar graphs are for the Cross Emergence Functions (CEFs). These are calculated for the probability distributions for the Source and Reservoir. The CEFs are constructed from the relevant ζ_0s and the respective cross entropies $H(p|r)$ and $H(r|p)$.

An important observation relates to the absence of the calculation of the CEFs between the Source-Sink in Figure 2.2. In this case, the cross entropies continue to diverge for the increasing ζ, thus indicating the growing cross entropic distances. Similarly, in the case of Reservoir-Sink combination, the pair of cross entropies showed low divergence and did not indicate stable emergent states as a function of ζ.

The CEFs for the Reservoir-Sink and Source-Sink combination cannot be evaluated because the respective cross entropies in Figure 2,2(b) and (c) continue to increase and do not reach stable maxima, as is the case for the Source-Reservoir pair. This is the consequence of the absence of the cumulative probability distributions in the example of the Source-Reservoir-Sink. Due to the conditions imposed on the present model, the emerging distributions in 2.1(a) to (c) follow the restrictions imposed by the condition of conservation discussed in the previous section as $p(\zeta) + r(\zeta) + q(\zeta) = 1$. However, in the case of the growing populations of the infected cases of Coronavirus pandemic, the cumulative probability distributions will not be restricted by the conservation condition and hence the resulting entropic and cross entropic profiles will demonstrate their respective maxima.

The Index of Divergence $\Delta\mathcal{E}$ is calculated from the absolute value of the difference of the two Cross Emergence Functions for the set of the twin distributions $p(\zeta)$ and $r(\zeta)$, as $\Delta\mathcal{E} = |\mathcal{E}(r|p) - \mathcal{E}(p|r)| \sim 157$ using Eq. (2.4).

The Reservoir to Sink cross entropy $H(r|q)$ demonstrates 3 times increase for the two information sharing stages $\zeta=20$ to $\zeta=40$. The Sink to Reservoir $H(q|r)$ shows a higher ratio of factor 5 for the two ζ ranges. This is due to the shrinking and the extinction of the Reservoir, and the consequent emergence and the buildup of the Sink.

The Cross Emergence Functions between the interacting constituents of the emerging, information generating dynamical systems provide the route of the emergence of the new states of the system. These CEFs yield the Indices of Divergence that may indicate the relative strengths of the emerging trends and the information-theoretic arguments for the evidence of the surviving and the evolving species. These will be shown to identify the pertinent features of the system under transition. In Chapter Seven, the evolution of the variants of SARS-CoV-2 will be identified through the diagnostic tools developed here: the Emergence Functions, Cross Emergence Functions, and the Indices of Divergence.

2.5. The Source-dominant SRS. The Source-dominant SRS has a Reservoir programmed to release the information received from the Source at a faster rate to the Sink. The Emergence Functions of the Source-dominant model were plotted in Figure 1.9 of Chapter One. Figure 2.3 has the sets of the cross entropies plotted as a function of ζ, for a similar Source-dominant SRS.

Figure 2.3(a) plots the profiles of the cumulative entropies evaluated for the information transfer from Source to Reservoir @1/32 with the subsequent Reservoir-Sink rate α =1/4, for the three probability distribution functions $p(\zeta)$, $r(\zeta)$ and $q(\zeta)$. The Source generates significant information as compared with the Reservoir which operates as the facilitator and the fast-transmission route to the Sink. The Source generates higher entropy than the Sink and the Reservoir, as Figure 1.9 of Chapter One showed.

The significant features of Figure 2.3(a) are that the two approximately equal entropic regions emerge; the first is the low ζ region for $\zeta \lesssim 15$ where the set of the three cumulative Shannon entropies have approximate equivalence $H(Source) \sim H(Reservoir) \sim H(Sink)$. For the region $\zeta \sim 15 - 50$, the entropies of the Source and the Sink continue to have an extended region of approximately equal entropies $H(Source) \sim H(Sink)$. The second ζ-regime is where $H(Source)$'s dominance over $H(Sink)$ and $H(Reservoir)$ is pronounced for $\zeta > 50$.

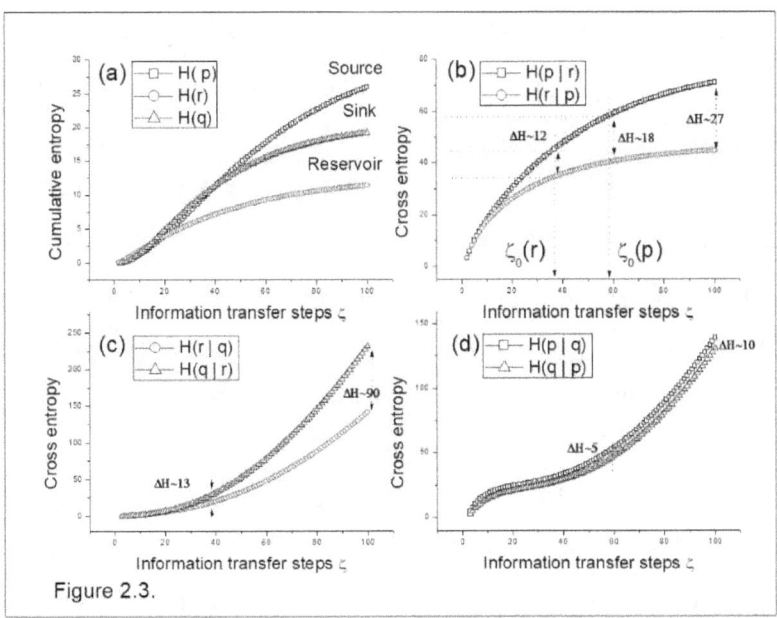

Figure 2.3.

Figure 2.3. The Source dominated DS. (a) The cumulative entropies for the three probability distributions are plotted as function of ζ. The Source shares its contents @1/32 per step ζ with the Reservoir and the Reservoir shares @1/4 with the Sink. From (b) to (d) the profiles of the three sets of the six cross entropies are plotted. (b) The Source-Reservoir set of the cross entropies $H(p|r)$ and $H(r|p)$ stabilize around constant values. (c) The set $H(r|q)$ and $H(q|r)$ generated by the Reservoir and Sink distributions demonstrate the continuously increasing cross entropies. (d) The cross entropies for the Source-Sink $H(p|q) \sim H(q|p)$ for the entire range of ζ, except for the first few steps where $H(p|q)$ is nominally larger than $H(q|p)$.

Three sets of the six mutual cross entropies calculated from the three probability distribution functions are plotted in Figure 2.3(b) to (d). The Source-Reservoir generated set of the two cross entropies $H(p|r) = \sum p(\zeta) \ln(1/r(\zeta))$ and $H(r|p) = \sum r(\zeta) \ln(1/p(\zeta))$ are shown as Figure 2.3(b). The region of equal cumulative entropies from $\zeta \sim 0$ to 15 in 2.3(a), is also represented as the region of equal cross entropies in 2.3(b). Beyond $\zeta \gtrsim 15$, the Source to Reservoir cross entropy shows the increasing inequality $H(p|r) > H(r|p)$. Both reach constant values of 75 and 45, respectively for $\zeta \gtrsim 100$. The difference between the cross entropies $\Delta H = [H(p|r) - H(r|p)] \sim 12$ at $\zeta_0(r)$ the inflexion point of the Reservoir and ~18 at the Source's inflexion point $\zeta_0(p)$. The divergence grows to ~27 for $\zeta \sim 100$. This is the case of a dominant Source and a compliant Reservoir.

Figure 2.3(c) has the twin cross entropies of the Reservoir and Sink $H(r|q)$ and $H(q|r)$. Notice that the y-coordinate scale is 10 times larger than the corresponding one in 2.3(b). The larger scale of the cross entropies implies the larger difference in the cross-entropy generation for the Reservoir and Sink distributions. The Sink-to-Reservoir cross entropy $H(q|r)$ is constantly increasing and getting larger than $H(r|q)$ symbolize the dominance of the Sink over Reservoir like the cumulative entropic profiles in Figure 2.3(a) where $H(Source) > H(Sink) > H(Reservoir)$. The magnitude of divergence is dependent on the difference between the rates of transmission between the Source-to-Reservoir and the Reservoir-to-Sink and

$\Delta H = [H(q|r) - H(r|q)]$ varies from ~13 at ζ~33 to ΔH ~90 at ζ~100.

Figure 2.3(d) plots the increasing cross entropies $H(p|q)$ and $H(q|p)$ for the Source-Sink combination. The two cross entropies have similar increasing trends with small divergence ΔH ~5 at ζ~33 to ΔH ~10 at ζ~100 This is the further evidence of the Source-dominated dynamical system where the Reservoir plays the role of the facilitator of the flow of information from Source toward the Sink, without making a significant difference. Here, the Reservoir plays the role of facilitator of the flow of information, but the Sink is dominantly influenced by the Source. The Source to Reservoir set of cross entropies produce a ratio $H(p|r)/H(r|p)$ that starts from ~1 and reaches to ~1.5 and stays constant after ~50 steps. The Source-to-Sink cross entropic ratio $H(p|q)/H(q|p)$ stays ~1 with throughout the transition for the state of the system transforming from {1,0,0} to {0,0,1}. That serves as an indicator of the Source-dominant dynamical system with small $\Delta H = H(p|q) - H(q|p)$~ 5 to 10, as shown in Figure 2.2(d).

The same argument can be provided for the absence of the cross-entropic maxima in 2.3(c) and (d) that was given for the similar cases in section 2.4. It is due to the nature of the distributions in the conservation conditions of $p(\zeta) + r(\zeta) + q(\zeta) = 1$. The condition restricts the Logistic growth of the Source and Reservoir populations. The present model deals with the conservative flow requirements.

2.6. The Reservoir-dominant SRS. If the rates of flow used to derive the dynamical system of Figure 2.3 above, are reversed, then from the Sink's perspective, the Reservoir will emerge as the 'effective' information generator, replacing the Source as described above. In Figure 2.4(a), the Source drains @1/4 to the Reservoir that shares its contents @1/32 with the Sink. The resulting cumulative entropies for the probability distributions $p(\zeta)$, $r(\zeta)$ and $q(\zeta)$ are shown in 2.4(a).

In Figure 2.4(b), the cross entropic profiles of the Source-Reservoir show that $H(p|r)) < H(r|p))$ within ~15 steps. The Source still recognizes the Reservoir by generating comparatively smaller cross entropy, but the Reservoir demonstrates a larger cross entropic divergence towards the Source as $\Delta H = [H(r|p) - H(p|r)]$ varies from ~0 for the first 10 information transfer steps to ~124 at $\zeta_0(r)$.

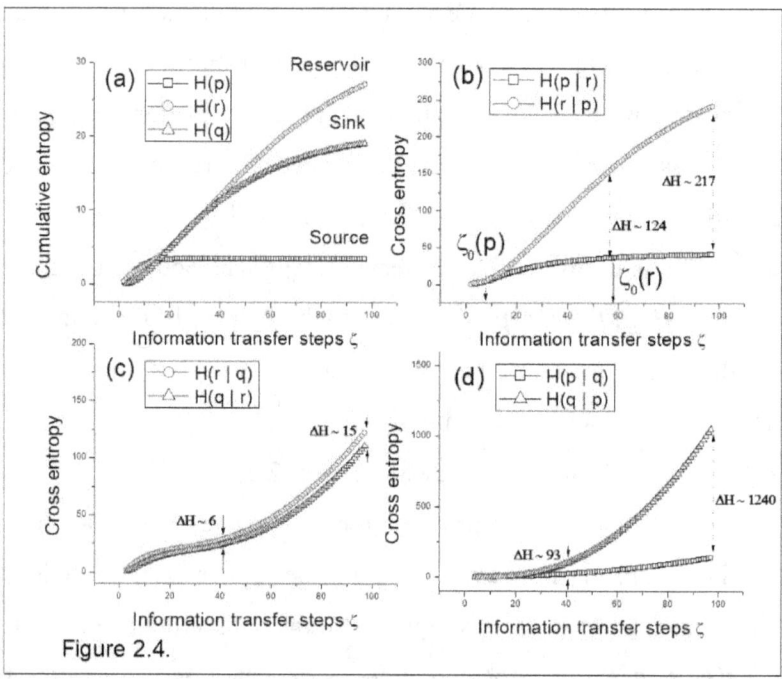

Figure 2.4.

Figure 2.4. The Reservoir-dominated DS. (a) The Cumulative entropies for the probability distributions $p(\zeta), r(\zeta)$ and $q(\zeta)$ are shown. The Source shares its contents @1/4 per step ζ with the Reservoir. The Reservoir shares @1/32 with the Sink. (b) The cross entropies $H(p|r))$ and $H(r|p))$ have been plotted against ζ. $H(r|p))$ is > than $H(p|r))$ after the first 15 steps. (c) The Reservoir and Sink generate $H(r|q))$~$H(q|r))$ for the entire information transfer process from $\zeta = 0$ to 100. This is the cross entropic evidence of the Reservoir-dominance on the information sharing dynamical system SRS. (d) The diverging cross entropies $H(p|q))$ from $H(q|p))$ further illustrates the diminished role of the Source for not being the effective Source.

Figure 2.4(c) has the almost equally distant cross entropies for the Reservoir–Sink combination that have $H(r|q)) \approx H(r|p))$. The Reservoir and the Sink continue to generate almost similar cross entropies up to $\zeta \sim 15$ steps. The maximum $\Delta H = [H(r|q) - H(q|r)]$ is ~15 at $\zeta = 100$.

The dominant role of the Reservoir, which operates as the de-facto Source for the Sink, is further elaborated by the diminishing cross entropy $H(p|q)$) as compared with the associated $H(q|p)$) in 2.4(d). Note that the difference $\Delta H \sim 93$ at $\zeta \sim 40$ steps and $\Delta H \sim 1240$ at $\zeta \sim 100$.

The SRS operates as the Reservoir-dominated dynamical system when the rate of flow from Source to Reservoir is >> than the corresponding rate from the Reservoir to Sink. The Reservoir holds the flow and the related information for much longer times after receiving it from the Source.

2.7. The Comparison of λs and CEFs of the three SRS categories. The Emergence and Cross Emergence Functions for the three types of information sharing and manipulating dynamical systems (Free-flow, Source-dominated and Reservoir-dominated) are plotted as bar graphs in Figure 2.5. The Emergence Functions λs are plotted for the Source, Reservoir, and the Sink for the three SRS systems in Figure 2.5(a) to (c). The random, Free-flow regime in Figure 2.5(a) projects the Reservoir with the largest λ followed by the Sink and the Source $\lambda(r(\zeta)) > \lambda(q(\zeta)) > \lambda(p(\zeta))$. The Source-dominated regime has the order reversed as $\lambda(p(\zeta)) > \lambda(q(\zeta)) > \lambda(r(\zeta))$ in Figure 2.5(b). The order of magnitude of the λs has also increased by a factor of ~10 as compared with the values in Figure 2.5(a). Figure 2.5(c) has the Reservoir generating the maximum λ. The decreasing trend of the λs is $\lambda(r(\zeta)) > \lambda(q(\zeta)) > \lambda(p(\zeta))$. The noticeable feature is the diminished numerical value of the Emergence Function for the Source $\lambda(p(\zeta)) \sim 25$, while the numerical value of the Reservoir's $\lambda(r(\zeta)) \sim 1974$. In the case of the Source-dominance in Figure 2.5(b), the Emergence Function $\lambda(p(\zeta)) \sim 1818$ and $\lambda(r(\zeta)) \sim 483$. This introduces the basic fact of the 3-component DSs that the role of the Reservoir is crucial to the functioning of the system, be it the Free-flow,

Source-dominated or the Reservoir-dominated regimes. The $\lambda(r(\zeta))$ for the Reservoir can reduce but it cannot be diminished like the $\lambda(p(\zeta))$ of the Source in Figure 2.5(c).

The Free-flow regime in Figures 2.5(d) plots the CEF bar graphs that have the ratio $\mathcal{E}(r|p)/\mathcal{E}(p|r) \sim 2$. The corresponding Index of Divergence $\Delta\mathcal{E} = |\mathcal{E}(r|p) - \mathcal{E}(p|r)| \sim 157$.

Figure 2.5(e) plots the Cross Emergence Functions bar graphs of the Source-dominant SRS, the CEF $\mathcal{E}(p|r) = 2958$ and $\mathcal{E}(r|p) = 1972$, with the mutual ratio $\mathcal{E}(p|r)/\mathcal{E}(r|p) \sim 1.5$. It generated the Index of Divergence between the Source and Reservoir $\Delta\mathcal{E} = |\mathcal{E}(p|r) - \mathcal{E}(r|p)| \sim 986$.

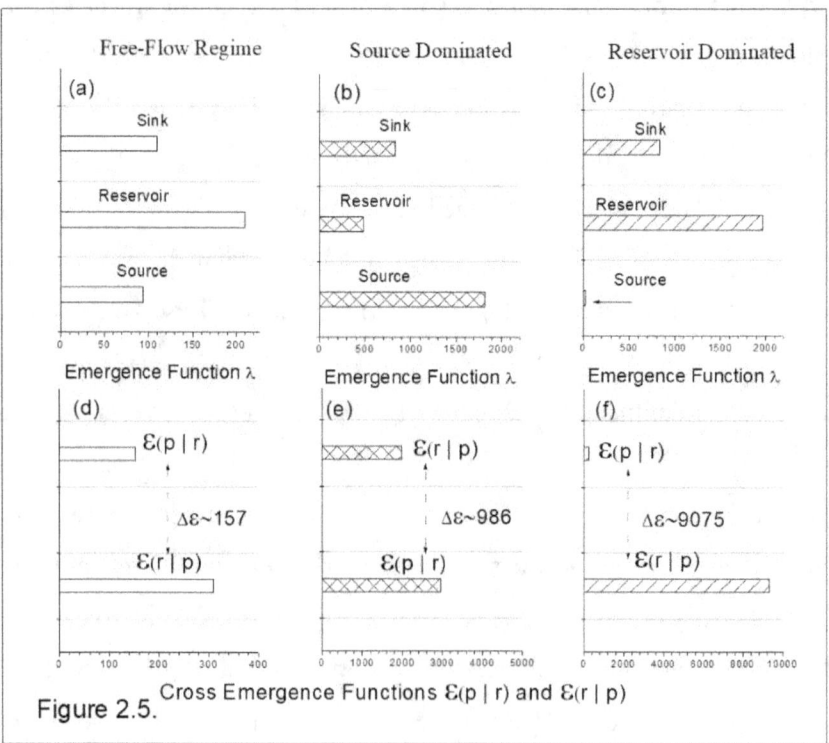

Figure 2.5.

Figure 2.5. The Emergence and Cross Emergence Functions. The Emergence Functions λs are plotted for the three configurations of the Source-Reservoir-Sink: in (a) for the Free-flow regime, (b) the Source-dominant, and (c) the Reservoir-dominant regimes. (d) The CEF for the Free-flow regime are plotted as bar graphs of $\mathcal{E}(p|r)$ and $\mathcal{E}(r|p)$. The $\mathcal{E}(r|p)$ is the dominant CEF that makes the larger contributions with the Index of

Divergence $\Delta\mathcal{E}$ ~157. (e) The CEFs for the Source dominated SRS has $\mathcal{E}(p|r) > \mathcal{E}(r|p)$ with $\Delta\mathcal{E}$ ~986. (f) The Reservoir dominated regime has $\mathcal{E}(r|p) > \mathcal{E}(p|r)$ with $\Delta\mathcal{E}$ ~9075.

The Reservoir-dominant SRS in Figure 2.5(f) is numerically described though the two CEFs: the $\mathcal{E}(r|p) = 9338$ that is greater than the corresponding $\mathcal{E}(p|r) = 263$ by a factor of ~36. The inequality $\mathcal{E}(r|p) \gg \mathcal{E}(p|r)$ implies that the Reservoir generates massively divergent cross entropy with the Source in this configuration where the Reservoir receives information at a much faster rate and delivers it to the Sink at a significantly slow rate. The buildup of the Reservoir as the dominant player in the information sharing dynamical system of the SRS, is due to the unequal rates that favor the Reservoir. The Reservoir-dominance implies the emergence of the Reservoir as the manipulator of the information received from the Source; such that the Source is considered distant from the Sink with massive cross-emergent divergence.

2.8. Summary. The relevance and applicability of the Emergence Functional technique for analyzing and characterizing entropy generation, sharing, and manipulation is extended in this chapter by adding the Cross Emergence Functions CEFs. CEFs use cross entropies generated by the comparable constituents of the dynamical systems. The chapter compared the characteristic features of the Logistic growths of the cumulative entropy, and cross entropy regimes in the emergence and cross-emergence. The essential features of the extended model graphically shown in the Graphical Summary are:

- The initiation of information-sharing by dynamical systems begins with the emerging profile of the Source of the 'original message.' The Reservoir manipulates the input before sharing it with the Sink. The key component of the entropic paradigm here is the implicit maneuverability of the 'original message.' This feature is the built-in characteristic of the probability distributions (a) and (b), for the Source-dominated and the Reservoir-dominated SRSs.

- Emergence Function λ and the Cross Emergence Functions are constructed from the emergence-defining parameters ζ_0 and H_m in the case of λ and maxima of the relevant cross entropies.
- The cross entropies evaluated for the probability distributions of the three pairs of the SRS constituents Source$\rightleftharpoons$Reservoir, Reservoir$\rightleftharpoons$Sink and Source$\rightleftharpoons$Sink. The intra-constituent entropic dynamics is shown as the ratios of the cross entropies in (c) and (d). The ratios of the six cross entropies $H(p|r)/H(r|p)$, $H(r|q)/H(q|r)$ and $H(p|q)/H(q|p)$ and are shown the Source-dominated and the Reservoir-dominated dynamical systems.

Graphical Summary has the bar graphs for the Cross Emergence Function-CEFs in (e) and (f). These were shown as Figure 2.5(e) and (f). Here, these are shown to complete the graphical representation of the dynamic processes that generate the probability distributions shown as (a) and (b). CEFs provide the quantitative measures of the entropic divergence between the constituents of the dynamical system and an estimate of the relative importance, or the extent of the contribution, made by the constituents, under the constraints of the system like Source-domination or the Reservoir-domination.

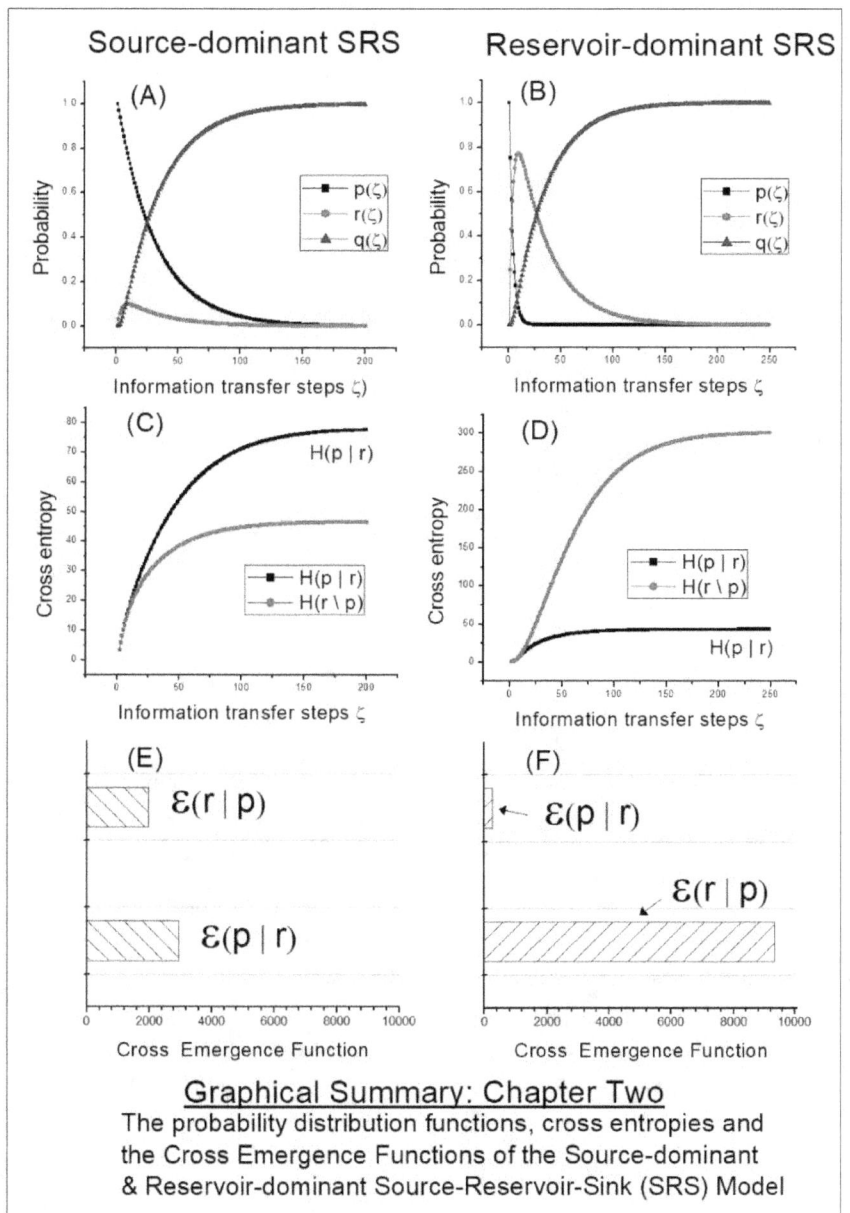

CHAPTER 3
PANDEMICS AS EMERGING DYNAMICAL SYSTEMS

3.1. Pandemic in 2002-4-the SARS CoV-2. Epidemics and pandemics can be treated as self-organizing dynamical systems. The infected person with the virus may become the repository and Source of virus transmission to others. The groups of the infected persons may become the Reservoir and the Source for further transmission and the initiation of a local epidemic. All epidemics possess the inherent possibility to cross the geographical boundaries and turn into pandemics. Information-theoretic interpretation of a pandemic is the transmission of the 'original message' at microscopic scales in few persons to the 'mega scale transmission' involving large populations. The transmission generate local as well as global entropic profiles. On the individual infected-person's scale, the spread of the virus in a living organism might reach to such a scale that the infected patient may not survive; dead body becoming the ultimate, yet local, Sink of the 'original message.' On the other hand, the Reservoirs that continue to grow with the increasing number of the cases of the infected persons, may merge with the others to facilitate the transition of the locally spreading virus to the global dimension of informational or the entropy generating pandemics. The 2002-4 data from the Severe Acute Respiratory Syndrome Coronavirus-2 categorized and designated as SARS-CoV-2 by the World Health Organization (WHO) is available from their website. It is in the form of daily situation reports of the new and the cumulative cases and deaths [70-71].

The SARS-CoV-2 pandemic was initiated in China's province of Guangdong. The first reported case occurred in Foshan, Guangdong on 16 November 2002. China officially informed the World Health Organization (WHO) on the 10th of February 2003 about the new disease of the respiratory system that had affected 305 cases that included 105 health workers and 5 deaths. A SARS-infected person Dr. Liu Jianlun from the Sun Yet Sen Hospital in Guangdong, travelled to Hong Kong. After falling sick, when admitted to Kwong Wah Hospital, he informed the staff about the contagious disease of the respiratory system that had spread in Guangdong. He died on 4th March 2003, by that date, he had become a super-spreader of the SARS-CoV-2, infecting 23 guests of the hotel where he stayed, who in turn took the virus to their countries and homes and functioned as the link-pins of the pandemic.

Approximately ~80% of the total of Hong Kong cases (~1755), were due to Dr. Liu [72-79]. The report on the Summary of the SARS cases from 1 November 2002 to 31 July 2003 is presented as Table 3.1 [70-71].

The WHO designated SARS-CoV-2 as a member of the Coronavirus family. An active virus from an uncertain animal reservoir, which had spread to other animals. The entire virus is surrounded by glycoprotein in the form of a corona or crown. The membrane and protein overlap, surround a genome of single-stranded RNA. The WHO reported that during 2002-4, less than 10 thousand cases were reported from 29 countries. The list of countries with the representative WHO data of the cases is presented in Table 3.1. The bulk of the total cases was in five countries/regions: 5327 in China, 1755 in Hong Kong, 346 in Taiwan, 251 in Canada and 238 in Singapore. The list displays the data from 1st November 2002 to 31 July 2003. It is a representative summary of the pandemic and will be used in this Chapter.

3.2. The salient features of the SARS-CoV-2 data. Table 3.1 presents the data of 29 countries/regions. It displayed certain distinctive features of the pandemic.

- 93.5% of the cases (7571 out of the total of 8096) occurred in China, Hong Kong, Canada, and Singapore.

Table 3.1	Cumulative number of Cases and Deaths				
Areas	Female	Male	Total	Number of deaths	fatality ratio (%)
Australia	4	2	**6**	**0**	0
Canada	151	100	**251**	**43**	17
China	2674	2607	**5327**	**349**	7
Hong Kong	977	778	**1755**	**299**	17
Macao	0	1	**1**	**0**	0
Taiwan	218	128	**346**	**37**	11
France	1	6	**7**	**1**	14
Germany	4	5	**9**	**0**	0
India	0	3	**3**	**0**	0
Indonesia	0	2	**2**	**0**	0
Italy	1	3	**4**	**0**	0
Kuwait	1	0	**1**	**0**	0
Malaysia	1	4	**5**	**2**	40
Mongolia	8	1	**9**	**0**	0
New Zealand	1	0	**1**	**0**	0
Philippines	8	6	**14**	**2**	14
Republic of Ireland	0	1	**1**	**0**	0
Republic of Korea	0	3	**3**	**0**	0
Romania	0	1	**1**	**0**	0
Russian Federation	0	1	**1**	**0**	0
Singapore	161	77	**238**	**33**	14
South Africa	0	1	**1**	**1**	100
Spain	0	1	**1**	**0**	0
Sweden	3	2	**5**	**0**	0
Switzerland	0	1	**1**	**0**	0
Thailand	5	4	**9**	**2**	22
United Kingdom	2	2	**4**	**0**	0
United States	13	14	**27**	**0**	0
Viet Nam	39	24	**63**	**5**	8
Total			**8096**	**774**	**9.6**

- The ratio of the Chinese number of the cases were 65.8% of the total cases. However, the case fatality ratio was ~6.6%. Whereas the average case fatality ratio in Hong Kong, Canada, and Singapore was ~16.7%.

- On 10 February 2003, China notified the WHO the 305 cases and 5 deaths due to Coronavirus. Starting from 16 November 2002, it took 86 days for the growth of cumulative cases from 1 to 305. On 26 March 2003, China had 792 cases with 31 deaths after 130 days of the first reported case.

- On 26 March 2003, Hong Kong notified 316 cases and 5 deaths after 44 days of the arrival of Dr. Liu Jianlun from Guangdong.

- The similarities and the differences between the data from China and Hong Kong will be investigated by utilizing the diagnostic tools developed in the first two Chapters.

- Canada's 251 cases grew over a period of 93 days as opposed to the 238 cases in Singapore over the period of 61 days. There was a gap of 23 days in Canadian cases with no new cases. The possibility of the twin waves may have existed and will be investigated later in this Chapter. Although the number are too small for authoritative assessment.

- The information-theoretic diagnostic tools of the entropic, cross entropic and the logistic analyses will be employed to provide the quantitative and comparative assessment of the means and methods of spread of the pandemic in the diverse environmental situation of different countries.

- (8) SARS-CoV-2 had a total number of the infected persons <10k. It will also be investigated that whether it be possible to employ the same analytical tools across the increasing number of cases from 10^4 to 10^6 to 10^9, during the 2nd episode of the Coronavirus pandemic that occurred with multi-million cases. The answer lies in the analysis and comparison within and across the landscape of the emerging entropic and the cross entropic analyses.

- (9) It will be shown that the information-theoretic SRS tools can be used to evaluate the characteristic properties of the diverse stages of pandemics. It will also be demonstrated that such systems are scale-invariant. The entropic tools can be employed to investigate the emerging profiles of the dynamical systems from the micro-to-mega scales of pandemics.

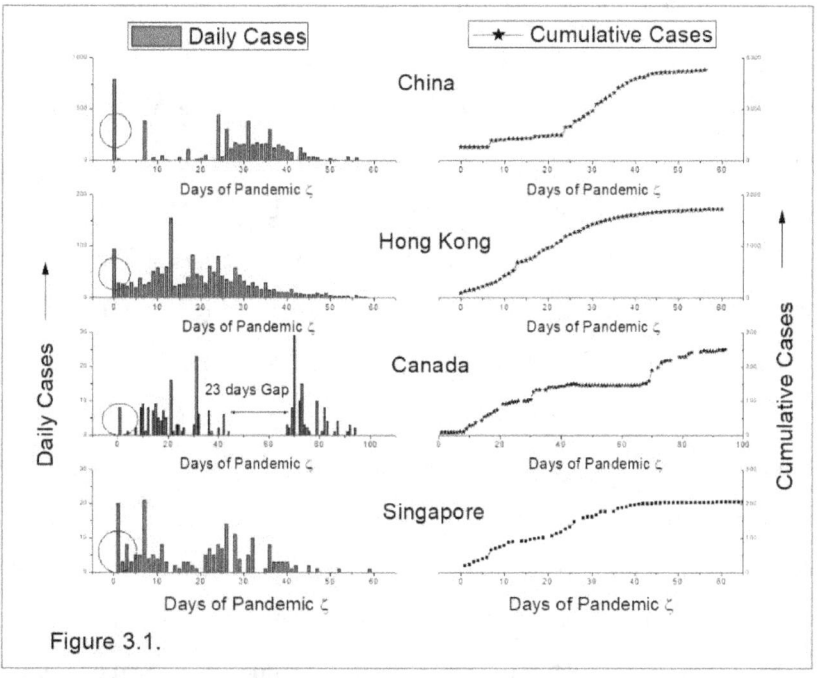

Figure 3.1.

Figure 3.1. The four countries/regions with ~93% of the SARS-CoV-2 Cases. Utilizing the data extracted from the WHO's daily situation reports [71], the daily and cumulative cases for China, Hong Kong, Canada, and Singapore are shown a function of the days of pandemic ζ. The cases constitute ~93.4% of the total cases. The disjointed nature of the histograms for the daily cases indicates the difficulty of identifying the initial dates of the pandemic. The cumulative cases show an emerging pandemic with the population of cases rising towards the maximum or the carrying capacity. Canada and Singapore had lesser numbers of the SARS-CoV-2 infected cases as compared with those in China and Hong Kong. Canadian graph for the daily cases clearly shows two episodes separated by ~23 days that produced the flat region between 40-70 days in the cumulative cases. Singapore data shows a continuously rising graphs of the cumulative cases. **Note: The earliest detected daily cases are encircled. These are the Source(s) carrying the 'original message' in terms of the information-theoretic paradigm, of the pandemic in the respective regions.**

3.3. SARS-CoV-2 as the Sources, Reservoirs and Sinks of the infected cases: Figure 3.1 is drawn from the data extracted from the WHO situation reports for the daily and the cumulative cases for China, Hong Kong, Singapore, and Canada [71]. The unifying feature of the daily cases in the four graphs is the spike of the initial virus infected persons at the starting point. These are encircled for identification of the onset of pandemic. The rates of increase and the inflexion points of the graphs for cumulative cases can be derived from the curve fitting using the Logistic equation of section 1.5 of Chapter One. It utilizes the initial number of cases on the first day of the pandemic q_0 in the derivation of each value of $q(\zeta)$ as function of the increasing number of the days of pandemic ζ. The Sigmoidal curve fitting of the four graphs of cumulative cases in Figure 3.1 yield the ζ value at the point of inflexion ζ_0, these are tabulated in Table 3.2(A). As the curve fitting has the initial values as the input parameter, lengthy delays and discontinuous reporting of the cases can affect the outcomes of the logistic values like the rate of increase and the exact values of the inflexion points ζ_0. The graphs of Figure 3.1 had their starting values q_0 shown encircled for China, and Hong Kong. The starting dates of the data have been indicated for each graph. The calculated inflexion points are reported in Table 3.2(A). However, in Figure 3.2 three graphs for the cumulative cases, one for Hong Kong and two for China are shown from the actual dates of the first few initial cases, instead of the large, encircled peaks of Figure 3.1. The Sigmoidal curve fitting now produces large divergences in the values of ζ_0s for Hong Kong and produces two values for China. These are reported as

Table 3.2(B).

Table 3.2(A) has the ζ_0 obtained by the Logistic curve fitting on the data for the cases reported for and presented as graphs in Figure 3.1. To add the impact of the initial reported cases in China and Hong Kong, the first reported cases are also added, for the Logistic curve fitting, and the new values of ζ_0 are tabulated in Table 3.2(B). The upgraded data is plotted in Figure 3.2. Hong Kong's first case emerged with the arrival of Dr. Liu on 21 Feb 2003 from Guangdong. The steady stream of the daily cases' data started from 26 Mar 2003, after a gap of 44 days. Figure 3.2(a) shows only the 105 days' data, fitted with the Logistic equation with $q_0=1$, resulting in 18.6 days in Figure 3.1. While the figure is yielding $\zeta_0=63$ days with $q_0=315$.

Similarly, in the case of the initiation of the Chinese pandemic, 16 Nov 2002 is designated when $q_0=1$. The first cases are encircled in Figure 3.2(a) and (b). China data, with the addition of the first reported case yields $\zeta_0=157$ days, as opposed to the $\zeta_0=41.3$ days reported for the 26 March to 26 May 2003 data, in Table 3.2(b). China's ζ_0s have three different values reported in Table 3.2(A) and (B).

The accuracy of ζ_0 from the data of the infected cases is important for the evaluation of the pandemic Emergence Function $\lambda = \zeta_0 * H_m$. The parameter λ will be used for the categorization and the comparative assessment of pandemic in different countries/regions. Therefore, another method of acquiring ζ_0 will be presented in the next section. It was pointed out in section 1.5 of Chapter One that the cumulative entropic graphs from the normalized probability distribution functions of the cases may yield the inflexion points which could overcome the problem of the logistic growth curves of the cases, as pointed above.

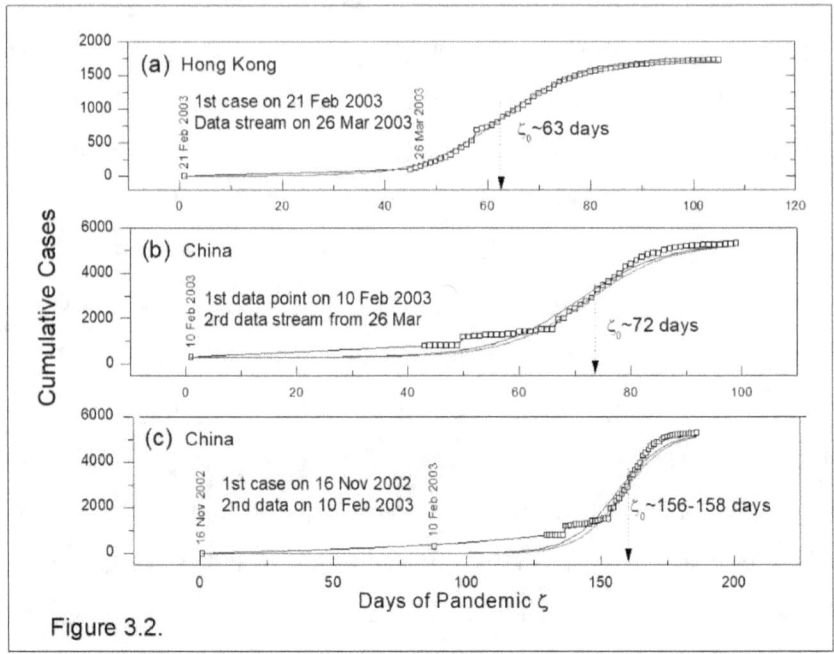

Figure 3.2.

Figure 3.2. The Logistic curve fitting of the upgraded emerging population of Hong Kong and China. (a) Logistic curve fitting is shown along with the data of the cumulative cases with the inclusion of the initiating Source of the SARS in Hong Kong starting from the case of the super-spreader arriving on 21 Feb 2003. The arrow points to the point of inflexion $\zeta_0 = 63$ days. (b) and (c) The Chinese data with the first reported case on 16 Nov 2002, and the other on 10 Feb 2023 are fitted with the two Logistic curves, shown by the thin continuous lines demonstrates the revised value of $\zeta_0 = 72$ days in (b) and 157-158 days in (c).

TABLE 3.2

(A) DATA FROM FIGURE 3.1	ζ_0 [DAYS]	(B) DATA FROM FIGURE 3.2	ζ_0 [DAYS]
CHINA	30	China	157
HONG KONG	19.3	China	72
CANADA	43.4	Hong Kong	63
SINGAPORE	19.7		

3.4. The SRS model applied to SARS-CoV-2. Two equivalent methods can be employed to characterize the Coronavirus pandemics. The first method is to evaluate the rates of the emerging populations of the cases by employing the Logistic equation and the second utilizes the information-theoretic emergence model of the SRS. The model developed and compared in the last two chapters investigates the nature of the emergence of the cases of viral infection in different situations as emerging, information generating, sharing, and manipulating dynamical systems. The Coronavirus can be treated as the self-organizing molecular system, the infected persons become the repository and reservoir of the virus for transmission to others. The application of the SRS model helps to characterize the dynamics of the emerging pandemic. The local versus the global emergence aspect of the pandemic will be identified using the SRS model in the next chapter while dealing with COVID-19.

The 7-days' incremental data for the growth of the population of cumulative cases is presented in Figure 3.3 for China and Hong Kong for the period of 26 Mar to 26 May 2003. The investigation is warranted to understand and clarify the clustering of data. After the detection of the Coronavirus in China and during the spreading of SARS-CoV-2 to Hong Kong, Taiwan, and Macau, the WHO issued situation reports showing the days without reporting of cases that resulted in the clustering that is evident in Figures 3.1 and 3.2. Figure 3.3(a) presents China's two months' data from the WHO situation reports [71]. The set of seven cumulative probability distribution functions $p(\zeta)$ is constructed by normalizing the data for each set starting from 14 days to 56 days at 7 days interval. Figure 3.3(d) has the same set of distributions for Hong Kong. The discontinuities in 3.3(a) can be contrasted with the relative smooth growing probability distribution graphs in 3.3(d).

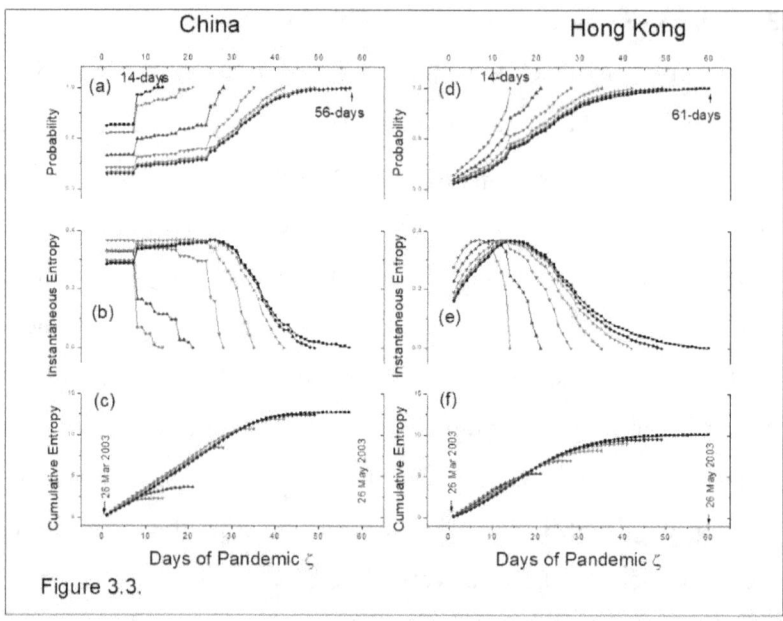

Figure 3.3.

Figure 3.3. The 7-days steps profiles for China and Hong Kong. The two-months of the pandemic between 26 March and 26 May 2003 are plotted as the cumulative probability distributions $p(\zeta)$, instantaneous entropy $p(\zeta)\ln(1/p(\zeta))$ and the cumulative entropy $\sum p(\zeta)\ln(1/p(\zeta))$ are plotted for China in (a)–(c) and for Hong Kong in (d)-(f). Each of the seven probability distribution functions is obtained by normalization of the data for the specific period.

Out of the seven-probability distribution $p(\zeta)$ graphs in Figure 3.3(a), the 3rd, 5th, and the 7th graphs yield the inflexion point ζ_0=24.2, 30.5 and 31.4 days. In Figure 3.3(d), the inflexion points ζ_0 varies from 10.4 days to 19.7 days increasing in 7 consecutive steps. The ζ_0 values for the whole duration 26 Mar to 26 May 2003 for China and Hong Kong match with the values obtained from the cumulative cases in Table 3.2(a). These normalized, cumulative probability distribution functions are used to evaluate the instantaneous and cumulative entropies.

The instantaneous entropy $p(\zeta)\ln(1/p(\zeta))$ graphs are plotted in Figure 3.3(b) and ((e). These graphs further illustrate that Hong Kong's probability distribution $p(\zeta)$ functions in 3.3(d) generate coherent peaks as compared with the discontinuous ones from 3.3(a). The cumulative entropic $\sum p(\zeta)\ln(1/p(\zeta))$ graphs for the two sets of seven probability distribution functions are shown in 3.3(c) and (f). A significant aspect of these cumulative entropic graphs is that the rates of growth of the entropic curves match within 10% of the rates obtained from the respective probability distribution graphs. The highest value is 0.3 for the 14 days to 0.17 for the 56 days period. These rates of growth are obtained by fitting the probabilities and the entropic graphs with the Logistic curve fitting.

Figure 3.4 is plotted with the data used to plot the emerging, cumulative profiles of the cases in China and Hong Kong for the same period used in Figure 3.3. The figure provides the graphic evidence of the information generated for each successive stage of the pandemic in terms of the accumulating number of cases and the corresponding Emergence Function λ.

The Chinese cumulative entropic graphs for the 7-days intervals, in Figure 3.4(a), show a slow growth during the first two stages, followed by large jump at the 28th day, and then steadily increases for the remaining stages. Hong Kong's graphs show the continuous entropic profiles in 3.4(b). These effects are clearly noticeable in the probability graphs of Figure 3.3 (a) and (d) and similarly in the instantaneous entropic graphs in Figure 3.3(b) and (e).

The Emergence Functions λ of Figure 3.4(c) have been calculated by using the ζ_0s and H_m evaluated from the consecutive cumulative entropic curves in Figure 3.4(a) and (b). The λ bar graphs display the collective effects of the inflexion point and the maximum generated entropy at each successive stage for the two countries. In Figure 3.4(c), the Chinese Emergence Functions λ as a function of the days of pandemic, show a sharp increase for the λ(28 days)~25 to 106 followed by a gradual transition towards λ(56 days)~266. The Hong Kong Emergence Functions plotted in 3.4(d) display a gradual increase λ(14 days)~28 to λ(21 days)~49 to λ(28 days)~84 to λ(35 days)~117….to λ(56 days)~180.

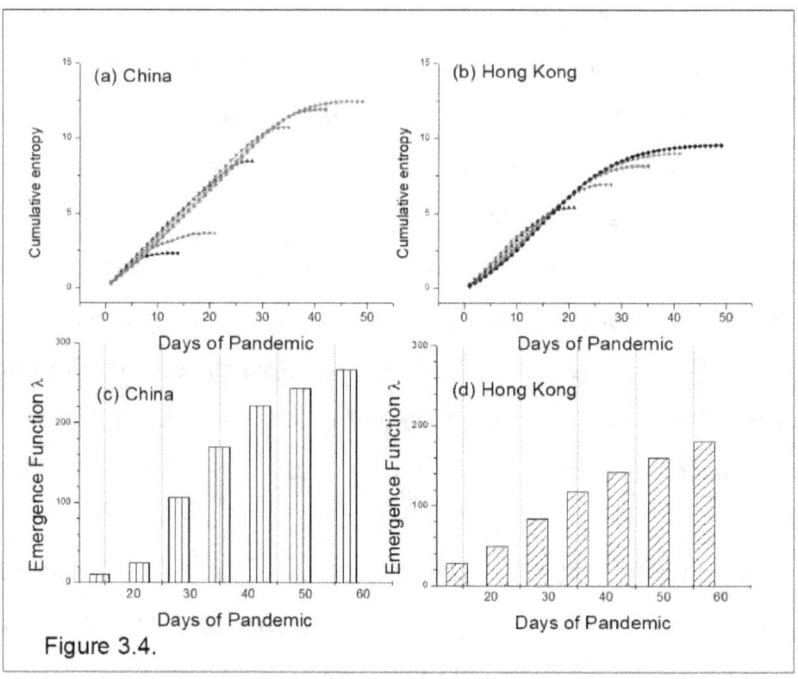

Figure 3.4.

Figure 3.4. The cumulative entropic and the Emergence Function graphs. (a) and (b) The cumulative entropic graphs for 7-days interval, starting from the 14th day, are plotted for China and Hong Kong. Figure (c) and (d) have the corresponding bar graphs for the Emergence Functions λs for each cumulative entropic curve in (a) and (b) as a function of the 7-days intervals of the days of pandemic.

The stage-wise evaluation of the Emergence Functional graphs can display the state of the pandemic and speed of the spread. However, the underlying assumption is the possibility to construct the cumulative entopic graphs at regular intervals. That condition may not always be fulfilled, as in the case of the Chinese data in Figure 3.3(a).

This procedure will be followed in the next chapter while dealing with COVID-19.

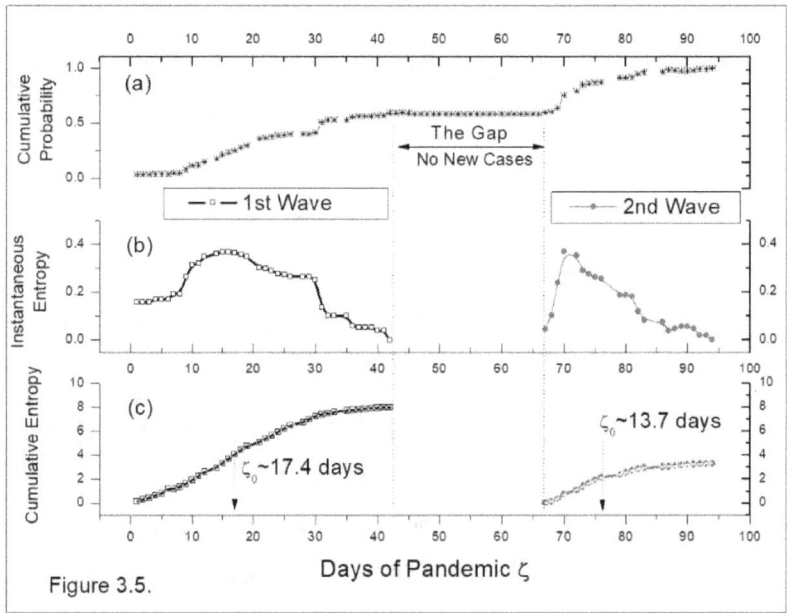

Figure 3.5.

Figure 3.5. The 2-wave analysis of the Canadian SARS CoV-2 data. (a) The cumulative cases' data is separated into the two sets of cases represented by the two distinct probability distribution functions $p_1(\zeta)$ and $p_2(\zeta)$. (b) The corresponding instantaneous entropies $p_1(\zeta)\ln(1/p_1(\zeta))$ and $p_2(\zeta)\ln(1/p_2(\zeta))$ for the two waves are plotted on the same horizontal ζ scale. (c) The cumulative entropies $\sum p_1(\zeta)\ln(1/p_1(\zeta))$ and $\sum p_2(\zeta)\ln(1/p_2(\zeta))$ are plotted for each wave with the inflexion points ζ_0s.

3.5. The possibility of the two-waves in the Canadian SARS-CoV-2.

Figure 3.1 for Canada can be further scrutinized to investigate the possibility of two discontinuous but related waves phenomenon. Figure 3.5 replots the data for Canada. It projects the separation of the cases' data as shown in Figure 3.5(a) around the 23-days gap into two sets: one with 142 cases and the other consisting of 103 cases. Each set of the separated cases is normalized to produce the two distinct probability distribution functions $p_1(\zeta)$ and $p_2(\zeta)$. These are used to derive the instantaneous entropies $p_1(\zeta)\ln(1/p_1(\zeta))$ and $p_2(\zeta)\ln(1/p_2(\zeta))$ shown in 3.5(b). The two waves of the cases are separated by the 23-days' gap. Figure 3.5(c) has the corresponding cumulative entropies $\sum p_1(\zeta)\ln(1/p_1(\zeta))$ and

$\sum p_2(\zeta)\ln(1/p_2(\zeta))$.

The sigmoidal curve fitting on the cumulative entropic graphs in Figure 3.5(c) yields the set of inflexion points ζ_0(1st wave)~17.4 and ζ_0(2nd wave)~13.7, as opposed to the value derived from the combined graph that included the 23 days of no new cases, in Figure 3.1 as $\zeta_0(combined)$~43.4 in Table 3.2(A).

The two derived Emergence Functions with the respective ζ_0s are λ(1st wave)=217 and λ(2nd wave)=132.

The ratio of λ(1st wave)/ λ(2nd wave)~1.64. The ratio for the Chinese and Hong Kong Emergence Functions λ(China)/ λ(Hong Kong)~1.48. The two ratios indicate that the second wave of the SARS-CoV-2 cases displayed smaller Emergence Function. The 2002-4 pandemic was diminishing and stopped spreading after the second waves, in Hong Kong and in Canada. The Canadian numbers are too small to conclusively illustrate this point.

The importance of the multi-wave phenomena will be dealt with in the next chapters that deal with the COVID-19 where the succeeding waves occurred with much larger ensembles of the virus infected cases. The next Chapter deals with the multiple wave phenomena, on the mega scale with hundreds of thousands and millions of cases. There the multi-wave phenomena will be elaborately demonstrated for the COVID-19 cases. The next chapters will highlight this fact. Here, the smaller scale, two-waves analysis demonstrates the dying stages of the pandemic.

3.6. The Cross Emergence Function. The cross entropic profiles are first calculated and then the Cross Emergence Functions (CEFs) of the probability distributions of the infected cases in China and Hong Kong be computed from these values of cross entropies plotted in Figures 3.6(a) and (b). The two waves of the cases in Canada are investigated by evaluating the cross entropies from the two normalized, cumulative probability distributions for the two consecutive waves that were separated by 23 days.

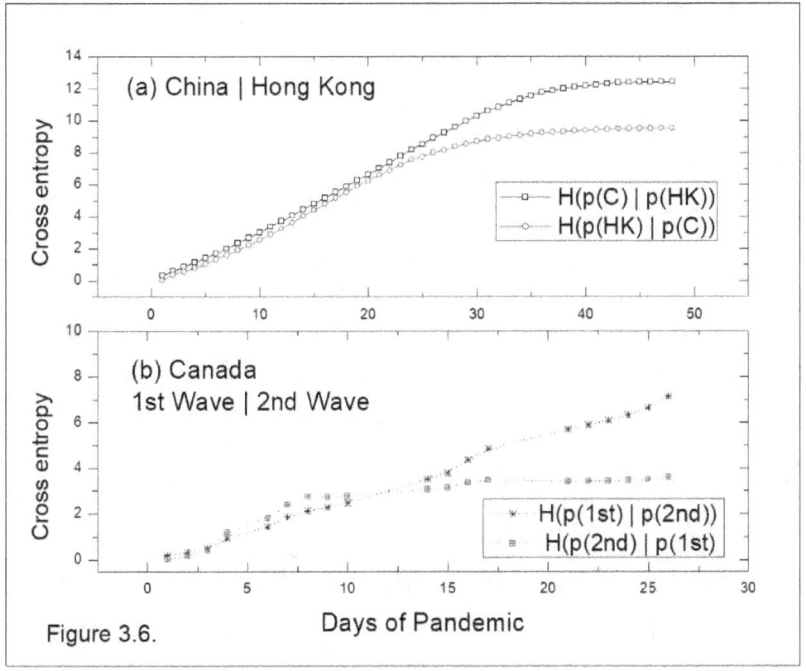

Figure 3.6.

Figure 3.6. The Cross entropies of China and Hong Kong and the 2 Canadian waves of cases. (a) The set of the twin cross entropies $H(p(C)|p(HK))$ and $H(p(HK)|p(C))$ for the Chines and Hong Kong cases' probability distribution functions $p(China) \equiv p(C)$ and $p(Hong\ Kong) \equiv p(HK)$ is plotted as a function of the days of the pandemic. (b) The cross entropies for the two successive waves of the Canadian cases $H(p(1st)|p(2nd))$ and $H(p(2nd)|p(1st))$ for the corresponding probability distribution functions $p(1st)$ and $p(2nd)$ is plotted as a function of the number of days of the pandemic.

The characteristics of the cross entropies evaluated and presented in Figure 3.6(a) and (b) can be summarized as follows.

In the case of China and Hong Kong, the cross entropies display small divergence in the graphs for $H(p(C)|p(HK))$ and $H(p(HK)|p(C))$. The cross entropies start to diverge for $\zeta \gtrsim \zeta_0(HK)$ as shown in Figure 3.6(a).

In the Canadian two-waves phenomenon in Figure 3.6(b), the two cross entropies $H(p(1st)|p(2nd))$ and $H(p(2nd)|p(1st))$ show relatively small divergence well beyond the inflexion points ζ_0(1st wave)~17.4 and ζ_0(2nd wave)~13.7

There is approximately equal net difference between the end points of the two sets of the graphs in the cross entropic units.

$\Delta[H(p(C)|p(HK)) - H(p(HK)|p(C))]$

$\sim \Delta[H(p(1st)|p(2nd)) - (p(2nd)|p(1st))] \sim 3$

The Cross Emergence Functions for China and Hong Kong can be evaluated by using Eq. (2.3) as

$\mathcal{E}(p(C)|p(HK)) = \zeta_0(p(C)) * H(p(C)|p(HK) = 373$, and similarly

$\mathcal{E}(p(HK)|p(C)) = \zeta_0(q(HK)) * H(p(HK)|p(C)) = 184$.

The Index of Divergence is the modulus of the difference of the two CEFs for China and Hong Kong is.

$\Delta\mathcal{E} = |\mathcal{E}(p(C)|p(HK)) - \mathcal{E}(p(HK)|p(C))| \sim 189$.

The Index of Divergence for the two waves of Canadian pandemic cases

$\Delta\mathcal{E} = |\mathcal{E}(p(1st)|p(2nd)) - \mathcal{E}(p(2nd)|p(1st))| \sim 77$.

3.7. Summary. The present chapter dealt with emergence of the first episode of the Coronavirus pandemic of 2002-4. Irrespective of the dwindling number of the cases~8K, reported by the WHO, the information-theoretic model presented in the first two chapters, was successfully employed to analyze the emergence and cross emergence of the SARS-CoV-2 as a dynamical system, with a global reach. The WHO reported 8096 cases in 29 countries/regions. Four of these 29 countries, China, Hong Kong, Canada, and Singapore reported ~90% of the cases. Therefore, this chapter focused on the applications of the information-theoretic SRS model to the dynamical characteristics of the spread of Coronavirus in these four countries/regions. The salient features revealed by the Emergence, Cross Emergence Functions, and the Index of Divergence of the 2002-4 episode with ~8K cases are:

- Pandemics can be treated as emerging dynamical systems. The information-theoretic model utilizing the information sharing and modulating Source-Reservoir-Sink configuration of the dynamical systems, has been used to characterize and evaluate the Coronavirus pandemic. By utilizing the entropic response and the logistic profile of the ensembles of the locally emerging pandemic cases, the SRS model employed the twin-component Emergence Function $\lambda = \zeta_0 * H_m$ to quantify the emerging features of the local pandemics in China, Hong Kong, Canada, and Singapore.

- The twin-dimensional Emergence Function is based on the holding capacity and the pace of the cumulative entropy generation by the growing ensembles of the accumulating cases of the pandemic. Together, the optimum value of entropy H_m and the pace of its growth determined by the inflexion point ζ_0 of the cumulative entopic curve, identify the information-generative profiles in Figures 3.4 through the Emergence Functional bar graphs for China and Hong Kong.

- Cross Emergence Function has been used as the qualitative and the quantitative measure of the cross-emergence of pandemics. The cross emergence and the inherent capability of the virus to spread in diverse groups of the local and global populations demonstrates its self-organizational character. This implies that the dynamical system adjusts itself according to the ambient conditions and adapts such that the transmission of the 'original message' be constantly modified and manipulated by the emerging Reservoirs of the virus infected persons.

- Cross Emergence Function helps to identify the 'selective' advantages of the virus in certain social groups, localities, countries etc. The twin cross entropic profiles, along with the inflexion points evaluated from the respective entropic graphs of the two distributions provide the quantitative estimates of mutual divergence through the dynamic dependence of the information-generating capabilities of each group.

- The CEF for the China-Hong Kong $\mathcal{E}(p(C)|p(HK))$ was shown to be twice as large compared with the CEF for Hong Kong-China $\mathcal{E}(p(HK)|p(C))$; demonstrating the extent of the divergence of the Chinese pandemic from the one in Hong Kong. Similarly, the CEF for the 1st wave in Canada versus the 2nd wave $\mathcal{E}(p(1st)|p(2nd))$ was ~2.5 times $\mathcal{E}(p(2nd)|p(1st))$. The initial or the parent ensemble of the population (China and the Canadian 1st Wave) had higher cross emergence as opposed to the daughter ensemble of the successor groups of the infected persons. That indicated a decaying or diminishing pandemic. That indeed happened; the pandemic lasted for about 100-150 days and stopped spreading. The Index of Divergence for the pandemic in China and Hong Kong $\Delta\mathcal{E}$ was ~2.5 the $\Delta\mathcal{E}$ for the 1st Wave versus the 2nd Wave in Canada.
- The evolutionary profile of the optimized emergence of SARS-CoV-2 in 2002-4 was tracked through the evaluation of the diversity of the pandemic by the Emergence Functions of each of the outbreaks. The selectivity of the virus's endurance was performed by the Cross Emergence Functions with the associated Indices of Divergence.
- The pandemic in 2002-4 demonstrated the profile of a dynamical system that could not self-organize against the further transmission restrictions into larger population groups. It did not display optimum productivity. The pandemic seem to be lacking the emerging features that are demonstrated through the diversity that the successor in 2019 displayed.
- The present chapter's conclusions regarding the lesser performance in 2002-4 will help to understand the optimized productivity, effective diverse global emergence, and the selection of the more efficient variants in cross emergence during the 2nd outbreak that started in 2019 and continues in 2024. That will be tackled in the next chapter. The 8×10^3 cases during $\sim 2 \times 10^2$ days in 2002-3 will be compared with the 8×10^8

cases during ~10^3 days of COVID-19 in Chapter Four.
- The deaths toll of the pandemic, local and global, are not considered in this book. The reasons are (a) the data for the emergence of cases has all the information needed to classify and diagnose the pandemic, (b) the cases of deaths are certainly important for the health care providers and the global community at large; however, the analyses will only bring out the differences in medical treatments in different countries and regions. That may not allow a unified approach to understanding the resulting deaths, and (c) the data of the infected cases are sufficient to present the emergence and cross emergence profiles of the pandemic. Our objective is to understand and analyze the pandemic as an emerging, self-organizing dynamical system.

The Graphical Summary highlights that the Emergence and the Cross Emergence Functions and the Index of Divergence can present the essential features and the evolutionary trails of the self-organizing pandemic in China and Hong Kong.

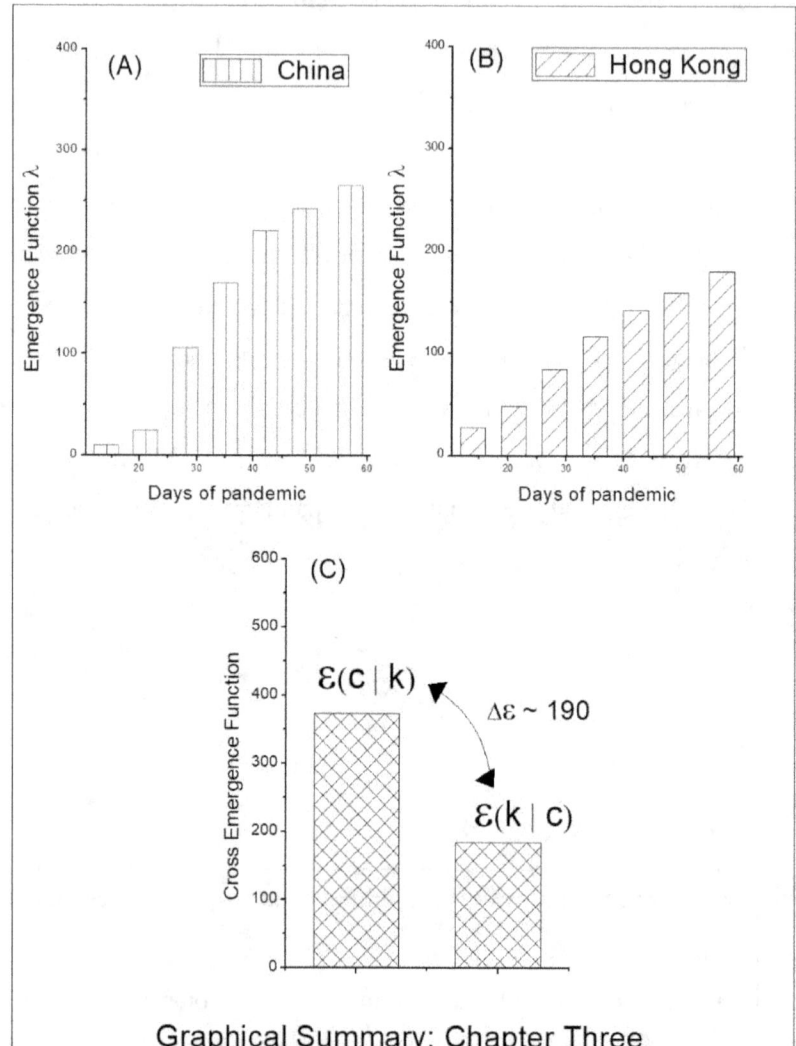

Graphical Summary: Chapter Three

<u>2002-4</u> SARS-CoV-2 pandemic in China and Hong Kong is shown through their Emergence Functions λ and the Cross Emergence Functions $\varepsilon(c \mid k)$ and $\varepsilon(k \mid c)$, using the SRS model of Chapters 1 & 2. The Index of Divergence between China and Hong Kong $\Delta\varepsilon \sim 190$

CHAPTER 4
THE EMERGENCE OF COVID-19

4.1. The 166 weeks of Emergence of COVID-19. The earlier episode of the Coronavirus pandemic in 2002-4, the SARS-CoV-2, was discussed in the last chapter. It now resembles the trailer of the real life, full-length, horror movie-the COVID-19. During 2002-4, about ~8.5k coronavirus-infected cases were recorded during the two years with less than a thousand deaths. Starting on the 31st of December 2019, with a few cases reported from China, by the end of 2020, the global total cases were ~90 million. The two years' total reached 308 million. The cumulative data of the 166 weeks is analyzed in this chapter that reports 757.3 million cases! It will be shown that the information-theoretic tools developed in the earlier chapters and used to evaluate the characteristic properties of the micro-scaled coronavirus pandemic in 2002-4, can be equally employed to the mega-scale COVID-19 data.

The duration of COVID-19 as 166 weeks, or 3 years and 10 weeks, represents the longest recorded and researched episode of a pandemic in modern history with ~760 million cases worldwide. We have based the analysis solely on the data provided by the WHO [80-87]. The WHO database is the repository of the global record of the cases, the fatalities, and the related details. At the time of writing of this book in early 2024, the WHO COVID-19 Dashboard is still active and adding the new data on daily and weekly basis. However, the existing data with three quarters of a billion cases, globally distributed, is believed to present a coherent and almost complete description of the pandemic.

The second important aspect of this book is the absence of the analysis and discussion of the approximate 1% deaths. This is a deliberate omission on the part of the author, and it was indicated at the end of Chapter Three.

There are multiple reasons; (a) the book is directed towards the global emergence of the pandemic, that will have its consequent deaths, and the short- and long-term effects, (b) the emerging pandemic is treated as a dynamical system that spreads in the context of the restrictions and controls imposed by the governments and public at large, the healthcare providing institutions, the local and international agencies, hence the data of the cases provide the emergence and cross emergent features of the Variants of Concern (VOCs) of the SARS-CoV-2, and (c) the data of the deaths and the fatalities will show the differences in gathering the statistics and the quality of the healthcare in different countries and regions.

Therefore, only the virus-infected cases are considered here. The population growth of the cases yields the dynamic emergence profiles of the pandemic at multiple levels. However, in Chapter 7, the data on the sequencing of the variants will be used for the identification of the Variants of Concern of SARS- CoV-2. That data includes all infected persons, those who survived and the ones who did not.

The emergent characteristics of the pandemic describe the dynamical systems that are scale invariant. This implies that the information-theoretic tools can be employed to investigate the emerging profiles of the dynamical systems that vary from the micro-to-mega scale. The emergence of self-organizing dynamical systems further demonstrates the sensitivity to initial conditions. The data from different regions, countries and environments provide ample evidence that slightly different initial conditions can generate widely differing outcomes of the pandemic.

The information-theoretic diagnostic tools of the cumulative entropic, and the cross entropic analyses will provide the quantitative and comparative assessment of the spread of the pandemic in similar or diverse environmental situation in different countries distributed in the six (6) WHO-designated regions of the world.

The earlier episode in 2002-4, of the SARS-CoV-2 had a total number of infected persons <10 k. We employ the same analytical tools across the increasing number of cases from 10^3 to 10^6 to 10^9! This aspect demonstrates the analytic capabilities of the information-theoretic tools to investigate the self-similarity and the scale-invariant dynamic character of the two episodes of the Coronavirus pandemics starting in the years 2002 and 2019.

Figure 4.1 presents the global data of the COVID cases from the six (6) WHO-designated regions. The figure presents the global and the regional weekly COVID-19 cases obtained from the WHO's web site https://covid19.who.int/ on the WHO Coronavirus (COVID-19) Dashboard [80]. The weekly data is for the duration of 166 weeks starting on 31st December 2019. It displays all the distinctive Pandemic features emerging during the entire duration of the 166 weeks.

The WHO data for the duration of 166 weeks is presented in the same form of the weekly cases as these appeared on the WHO Coronavirus Dashboard on its web site https://covid19.who.int/ [80]. The global total of the cases from 6-regions, is shown in Figure 4.1(a) as a function of the weeks of COVID-19. There are several smaller peaks arising from the consecutive waves of new cases in different regions and countries. Besides multiple smaller peaks, two massive peaks around the 110 and 155 weeks emerge over and above the high background of millions of cases. These peaks in the data represent the weekly sum of the global cases.

Figure 4.1(b) shows the profile of the emerging pandemic with a massive 273 million cases in Europe. The high and broad peak around the 110th week dominates the data. Figure 4.1(c) demonstrates a different emergence pattern in the Western Pacific region. The 201 million COVID-19 cases in Western Pacific region demonstrate a consistent pattern of smaller number of cases for the three years i.e., from weeks 0 to 155. The three years of lesser number of cases and multiple smaller peaks can be seen leading to the massive ~100 million cases' peak around the 155th week. The pattern and profile of Coronavirus spread in Western Pacific will be discussed in a later section.

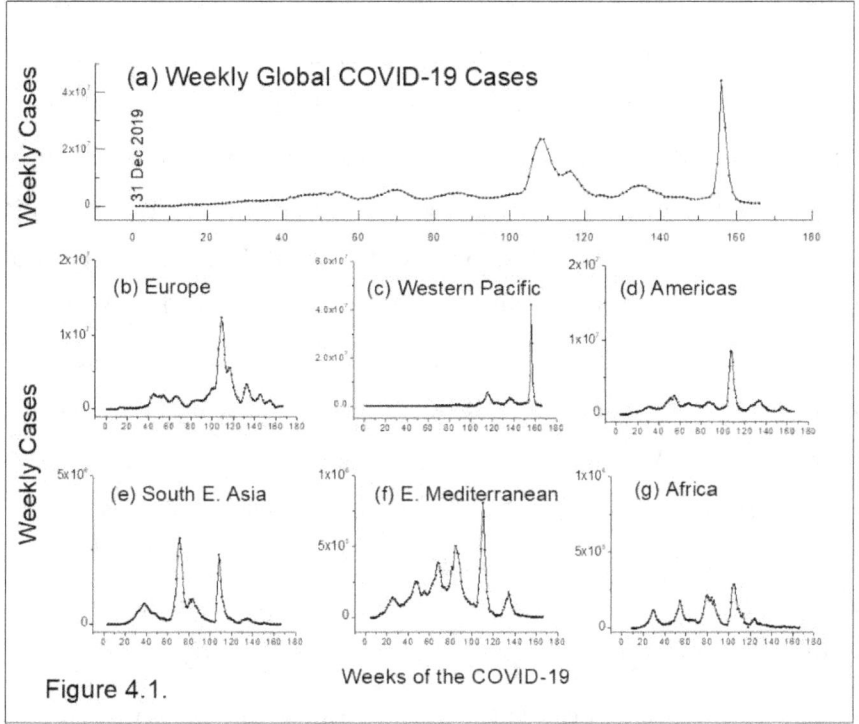

Figure 4.1.

Figure 4.1. The 166 weeks of the 757 million COVID-19 cases. (a) The weekly data from the WHO web site https://covid19.who.int/ for the total of the global cases from 6-regions, is shown a function of the weeks of the COVID-19. The starting date is indicated as 31 Dec 2019. Besides smaller peaks, two massive peaks around the 110th and 155th week are noticeable. (b) The profile of the 273million cases in Europe is shown. The high and broad peak around the 110th week dominates the data. (c) The emergence of 201million COVID-19 cases in Western Pacific region has multiple smaller peaks leading to the massive ~100 million cases' peak around the 155th week. (d) The data of 190 million cases for the Americas also displays the 110th week peak along with several smaller ones. (e) 60.8 million cases in Southeast Asia are spread around the 2 larger peaks at 70th and the 110th week. (f) 23 million cases in the East Mediterranean region have multiple peaks including the dominant one at 110th week. (g) 9.5 million African continent's cases show 4 broad peaks; the 110th week is also present. **The origin of the peaks, due to the emerging variants, will be discussed in Chapter 7.**

Figure 4.1(d) shows the data of the 190 million cases for the Americas region. It displays the 110th week peak along with the smaller ones. 4.1(e) has a total of 60.8 million cases in 166 weeks in Southeast Asia. These are spread around the two larger peaks at 70th and the 110th week. 4.1(f) shows the 23 million cases in the East Mediterranean region spread into multiple peaks including the dominant one at 110th week. 4.1(g) has the 9.5 million African continent's cases with 4 broad peaks; the 110th week is also present. Africa had the lowest number of cases among the 6 regions of the WHO.

The origin of the multiple waves/peaks will be analyzed using the data on the evolutionary trail of the SARS-CoV-2 by identifying the sequential emergence of the Variants of Concern in Chapter Seven.

It is important to note that the chaotic nature of the initial few cases are the essential proofs of the pandemic and may also determine the initiation as well as the final profile of the pandemic. The first Situation Report-1, by the WHO classified as 'the Novel Coronavirus (2019-nCoV)' was issued on 21 January 2020 [81]. It highlighted various aspects of the new virus. Its summary was: (1) On 31 December 2019, the WHO China Country Office was informed of cases of pneumonia of unknown etiology (unknown cause) detected in Wuhan City, Hubei Province of China. From 31 December 2019 through 3 January 2020, a total of 44 case-patients with pneumonia of unknown etiology were reported to WHO by the national authorities in China. During this reported period, the causal agent was not identified. (2) On 11 and 12 January 2020, WHO received further detailed information from the National Health Commission China that the outbreak is associated with exposures in one seafood market in Wuhan City. (3) The Chinese authorities identified a new type of Coronavirus, which was isolated on 7 January 2020. (4) On 12 January 2020, China shared the genetic sequence of the novel Coronavirus for countries to use in developing specific diagnostic kits. (5) On 13 January 2020, the Ministry of Public Health, Thailand reported the first imported case of lab-confirmed novel coronavirus (2019-nCoV) from Wuhan, Hubei Province, China. (6) On 15 January 2020, the Ministry of Health, Labor and Welfare, Japan (MHLW) reported an imported case of laboratory-confirmed 2019-novel coronavirus (2019-nCoV) from Wuhan, Hubei Province, China. (7) On 20 January 2020, National IHR Focal Point (NFP) for Republic of Korea reported the first case of novel coronavirus in the Republic of Korea.

The WHO later re-designated the 2019-novel coronavirus (2019-nCoV) as COVID-19, it will be referred to as such later in this Chapter and the succeeding ones.

The pandemic that led to three quarters of a Billion cases in just over three years, started with the few, initial carriers of the virus; some of these were the super-spreaders, like Dr. Liu in Hong Kong in early 2003 [71-75]. The next Chapter will investigate the initial 100 days of the COVID-19. Here, the cumulative 166 weeks' data of COVID-19 will be treated with the information-theoretic tools developed in the first two chapters.

The weekly growth provides a convenient order of magnitude estimation. From this data. the cumulative populations of the cases will be used to derive the normalized probability distribution functions for the individual regions and the global populations. The respective probability distribution functions of the cumulative cases yield the cumulative entropies and the cross entropies for the regions. The inter- and intra-regional comparisons will require the cross entropic estimates. Cross entropy will be used for comparative and quantitative evaluation of the emerging pandemic in different conditions, countries, and the WHO-designated regions.

The Emergence Function $\lambda = \zeta_0 * H_m$ defined in Chapter One, will serve as a distinctive property of the emergence, during and after the pandemic. The Emergence Function λ conveys the same information, as it did in the earlier chapters. Here, it is applicable to a broader range of the emerging scenarios and sequences, as will be demonstrated. The present chapter is focused on the emergence of the mega pandemic COVID-19; therefore, the emphasis is on identifying its patterns and profiles.

4.2. Global comparison of emerging population of the COVID-19 cases. The weekly global data from the WHO-designated regions presented separately in Figure 4.1 are shown in Figure 4.2 as the emerging, cumulative population of the coronavirus-infected cases. The six graphs represent the comparative emergence of the Coronavirus in the respective regions with certain similarities and the subtle differences. The significant feature is the percentile ratio of the 757.3m global cases in the regions of Europe, W Pacific, Americas, S E Asia, E Mediterranean, and Africa as **36:26.5:25:8:3:1.25**, implying that the 87.7% of the global cases emerged in the regions of Europe, Western Pacific and the Americas, the rest 12.3% of the cases constituted the S E Asian, E Mediterranean, and African share of the global Coronavirus cases.

Figure 4.2 is shown as sectioned in the two, boxed, early years. The other two vertically elongated, dotted rectangular boxes identify the emerging cases' profiles between the 100-120 weeks period and the 156±4 weeks, respectively. Europe provided the largest number of cases, 273m in 166 weeks; starting from 31 December 2019 and ending with ~30m cases in the first year. It had 112.6m cases at the end of the 2 years. Its total cases increased by 130m in the 100-120 weeks. This massive peak can also be identified in Figure 4.1(b).

The Coronavirus had emerged in one of the W Pacific countries, China. The region demonstrates a unique profile as can be seen in Figure 4.2. During the first 100 weeks of the pandemic, the region had ~10.5m cases. The cumulative cases increased by 53m in the next 30 weeks and another 31.8m cases after the next 20 weeks. During the entire period of 0-150 weeks, the region had a total of 95.3m cases. Then came the big surge of 105.7m cases within the 16-week period from the 150th to 166th weeks. The region, along with the others will be further analyzed in the next sections and Chapter 5.

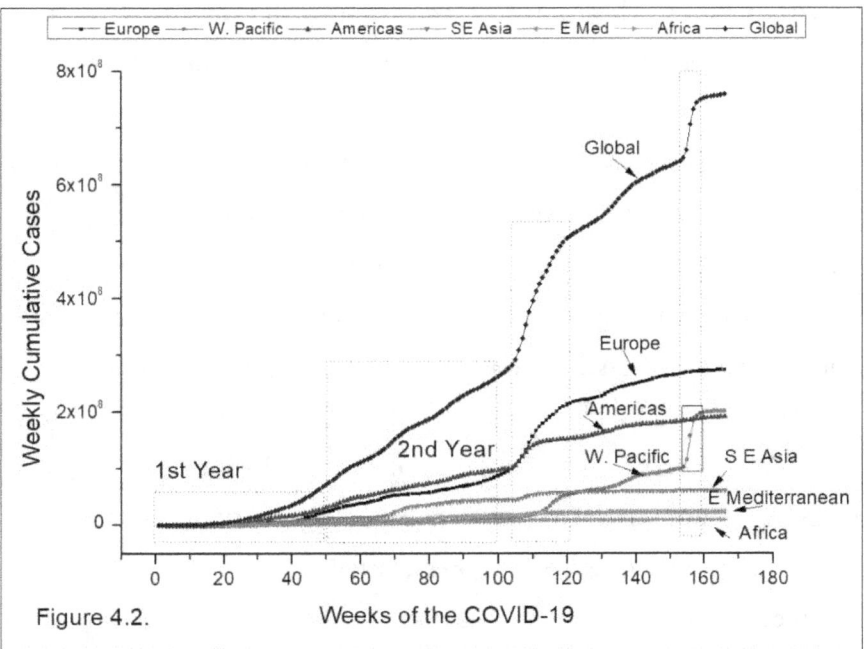

Figure 4.2.

Figure 4.2. The profile of the cumulative cases in the six(6) WHO-designated regions. The weekly data presented in Figure 4.1 for 166 weeks for the six regions is shown here as the cumulative cases as a function of the increasing weeks of the COVID-19. The 1st year, shown as the first rectangular box, added 90m cases. Out of the total of 90m, Europe had 30m, Americas 24.5m, S E Asia 12.3m. W Pacific had 1.2m cases in the first year. The second year ended with 308m cumulative cases in 2 years with Europe and Americas contributing ~ 224m cases, the S E Asian region had 48m, W Pacific with 13m and E Mediterranean had 17.5m cases. Africa contributed only 7.6m during the 2 years. Between 100 and 120 weeks, massive surge occurred in all but the E Mediterranean and the African regions. During the period of the 100-120 weeks, Europe increased its total of the 2 years' cases by 130m, Americas by 55m, W Pacific by 42m and S E Asia 8.2m. Another massive surge of ~100m cases occurred only in the W Pacific region between the period 152-160 weeks. This was the lone, W Pacific surge in the 4th year, the unique mega event that was not observed elsewhere.

With 190.8m coronavirus cases, the Americas region had the third largest population of cases at the end of the 166th week. At the end of the first year there were 39.2m cases. The ongoing Coronavirus increased its total to 111.75m at the end of the second year. The period 100-120 weeks produced another 55m cases. This increase can be identified in Figure 4.2 as the sharp upward increase in the cumulative cases graph and a singular peak in Figure 4.1(d).

S E Asian graph has the evidence of the three waves of high cases around the 40th, 70th and 110th weeks. The corresponding peaks can also be seen in the weekly cases in Figure 4.1(e). Coronavirus generated a total of 60.8m cases in the region. The footprints of the peak around 110 weeks are visible, as it can be observed in the European, W Pacific and the Americas.

E Mediterranean region demonstrated six consecutive waves in Figure 4.1(f). With the total of 23.3m cases, the regional cumulative graphs in Figure 4.2 shows the increasing cumulative cases profile. Its emergence profile will be more clearly visible in the detailed analysis later.

African region had the lowest number of cases ~9.48m in 166 weeks. Its corresponding graph in Figure 4.1(g) shows four clear waves of the cases. These will be further discussed in the detailed regional analysis using the information theoretic Emergence Function λ as the basic tool.

4.3. Entropic profiles of the emerging COVID populations. Entropic profiles are constructed from the probability distribution functions for each set of the cumulative cases from the WHO-data as a function of the increasing number of weeks. The Shannon entropy from Eq.(1) of Chapter One implies that for the period 0-166 weeks, $H(\zeta = 0 - 166) = \sum_{\zeta=0}^{166} p(\zeta) \ln(1/p(\zeta))$. The probability distribution functions $p(\zeta)$ for each chosen range of ζ are constructed from the normalized data. For example, The first graph in Figure 4.3(a) is for the $\zeta = 0 - 20$ weeks. The probability distribution $p(\zeta)$ was normalized with the cumulative cases' total of 610817 cases. The graph denoted as (1) in Figure 4.3(a) represents the cumulative entropy constructed from the associated $p(\zeta)$. It is displayed as a function of ζ-the weeks of COVID-19. Similarly, all other entropic graphs for the successively increasing ranges of ζ utilize the appropriately constructed $p(\zeta)$ for each specific range. The sequence of the entropic graphs is used to calculate ζ_0-the inflexion points where the holding capacity of the cumulative entropy is halved.

Figure 4.3(a) – (g) have the Shannon entropic graphs for the successive, distinct, emerging stages of the Coronavirus cases. Each entropic graph utilizes its probability distribution function $p(\zeta)$. (a) Global entropic graphs for the eight (8) stages of the evolving pandemic are shown with a graph for each stage. (b) European COVID-19 cases are distributed in six (6) distinct stages of the emerging pandemic. (c) Western Pacific region displays the unique distribution of the cases in the four (4) disjointed stages. The special emergent profile of the cases in the W Pacific region was pointed out in the previous section. It will be further elaborated in the next section that treats each region separately. (d) The Americas region demonstrate the six (6) stages of emergence. The S E Asian, E Mediterranean and African entropic graphs, with their respective stages of emergence are shown as the three sets of graphs Figure 4.3(e) to (g).

These entropic graphs carry the cumulative information generated for each region with the constituent graphs for each separate stage of the evolving pandemic. These can be used to provide the following clues to the emergence of COVID-19.

How much information was generated - the quantitative aspect; and the pattern (duration, rate of growth of cases, etc.) of the spread of the Coronavirus.

Identification of the intra-stage (virus-spreading) virus-communication in terms of dependence on the earlier stages (to be analyzed in Chapter Seven). The intra-regional COVID-19 transmission profiles are built and compared by using cross entropic analyses. The cross entropic will be shown as a potent tool to investigate the intra-stage and the inter-regional emergence profiles of pandemic.

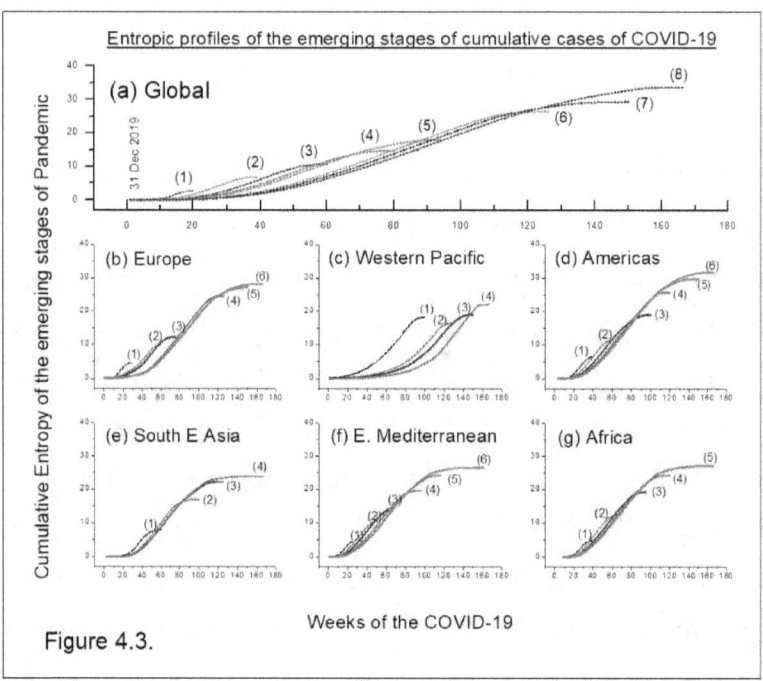

Figure 4.3. The cumulative entropic profiles for the successive, distinct, emerging stages. (a) Global entropic graphs for the eight (8) stages of the evolving pandemic are shown with a separate graph for each stage. (b) European COVID-19 cases are distributed in six (6) distinct stages of the emerging pandemic. (c) Western Pacific region displays unique distribution of the cases in the four (4) disjointed stages. (d) Americas demonstrate the six (6) stages of emergence. S E Asia, E Mediterranean and Africa are shown in (e) to (g) with the consequent stages indicated as graphs in each figure.

The above-mentioned features of the information-theoretic profiles generated by the entropies and the cross entropies, will illustrate the successive emergence and the sustenance of the Coronavirus observed globally and within the regions. In the next section the Emergence Function λ will be evaluated for the respective stages of the regions. The Emergence

Functions λs identify and distinguish the emerging profiles of each stage independently, and in the context of the global Coronavirus.

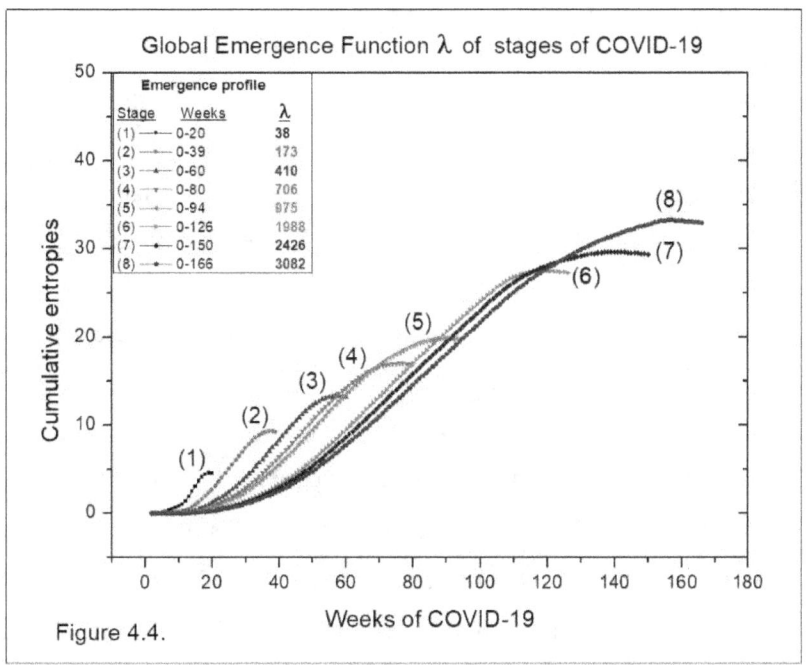

Figure 4.4.

Figure 4.4. The global Emergence Function of COVID-19 is evaluated from the cumulative entropic graphs for the eight(8) increasing stages of the pandemic. Emergence Function λ, for each stage, is calculated and shown in the boxed **inset**.

4.4. The global COVID-19 Emergence Function λ.

The set of graphs of the cumulative global entropies as a function of ζ-the weeks of COVID-19 are shown in Figure 4.4. The eight stages of the evolution of the global COVID cases are presented as individual graphs. Each successive stage includes the previous ones. Separate, distinct probability distribution functions $p(\zeta)$ are evaluated for each successive stage, by normalizing cumulative data for the stage. The cumulative entropic graphs for $(H(\zeta) = \sum_{\zeta=0}^{x} p(\zeta)\ln(1/p(\zeta))$ are used to evaluate the inflexion point ζ_0. The emergence function $\lambda = \zeta_0 * H_m$ is evaluated for each stage and shown in the boxed **inset,** with λ and its input parameters are separately tabulated as Table 4.1.

TABLE 4.1. GLOBAL

CUMULATIVE WEEKS	ζ_0	H_m	$\lambda = \zeta_0 * H_m$
0 – 20	14.7	2.69	38
0 - 39	25.44	6.78	172.5
0 – 60	37.97	10.8	409.6
0 – 80	47.7	14.8	706
0 – 94	54.35	17.93	974.5
0 – 126	75.3	26.4	1987.9
0 – 150	82.39	29.44	2425.6
0 - 166	91.53	33.67	3081.8

4.5. The European and American regions' COVID-19 Emergence Functions. Figure 4.5 presents the entropic graphs as a function of ζ-the weeks of COVID-19 for the two regions, Europe, and the Americas. The set of graphs of the cumulative entropies for each successive stage for Europe are shown in Figure 4.5(a). The six stages of the evolution of the COVID cases are plotted as individual graphs calculated from the six normalized, cumulative probability distribution functions constructed for each stage. Similarly, the set of six graphs of the successive entropic profiles for the countries in the region that spans the two American continents, are shown in Figure 4.5(b).

The similarities in the two regions' COVID profile of the entropic dimensional graphs had already been shown to emerge in the data for the weekly cases presented in Figure 4.1(b) and (d). Both regions display similar stages of the evolving Coronavirus cases. Though the total number of cases differ significantly, 273m cases in Europe versus 191m in the Americas, the overall pattern of the waves of the cases and the emerging stages present similar emergence profiles.

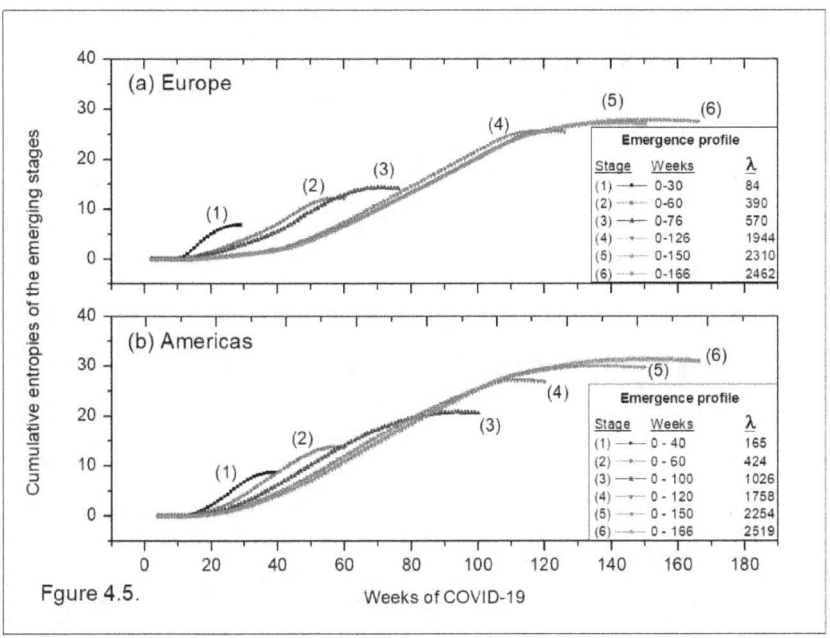

Figure 4.5. Cumulative entropies generated by COVID-19 in Europe and the Americas. The six (6) graphs of the cumulative entropies for (a) Europe and (b) the Americas, are plotted for the same stages that were shown in the corresponding weekly graphs in Figure 4.1(b) and (d). The entropic graphs for the corresponding stages of the evolution of the pandemic were calculated and shown in Figure 4.3(b) and (d). Here, the stage-wise, entropic profiles are replotted. These are used to calculate the Emergence function for each stage shown in the two boxed **insets**. The two sets of graphs are used to derive the Emergence profile by evaluating $\lambda = \zeta_0 * H_m$.

The cumulative entropic landscape as presented above in Figure 4.5 of the two of the world's most active regions where COVID-19 emerged and continued to spread in wave after waves, is tabulated below as Table 4.2 from the inflexion points ζ_0s, maximum generated entropies H_m and the resultant Emergence function λ.

4.6. The W. Pacific COVID-19 Emergence Functions. The set of graphs of the entropic graphs as a function of ζ-the weeks of COVID-19 in the countries on the rim of the W. Pacific region are shown in Figure 4.6(b). The four well recognizable stages of the evolution of the population

TABLE 4.2

CUMULATIVE WEEKS	ζ_0	H_m	$\lambda = \zeta_0 * H_m$
(A) EUROPE			
0 – 30	18.06	4.6	84
0 - 60	39.36	9.9	390
0 – 76	46.41	12.28	570
0 –125	78.7	24.7	1944
0 – 150	85.23	27.1	2310
0 – 166	87.46	28.15	2462
(B) AMERICAS			
0 – 40	25.95	6.37	165.3
0 - 60	37.85	11.2	423.9
0 – 100	54.0	19.0	1026
0 –120	68.14	25.8	1758
0 – 150	75.74	29.76	2254
0 – 166	79.2	31.8	2518.6

of COVID cases are presented as the four distinct graphs of H_m versus the weeks of COVID-19. These graphs are obtained by constructing the cumulative probability distribution functions for each of the four successively increasing period of the coronavirus pandemic. To emphasize the unique nature of the COVID-19 in the region, and for the sake of clarity, the weekly cases are also shown in 4.6(a) for the clear identification of the stages or the waves of cases. These weekly cases had already been included in Figure 4.1(c). The Shannon cumulative entropic graphs of the four emerging stages were shown in Figure 4.3(c). Here, in Figure 4.6(b), the entropic graphs once again confirm the unique aspects of the emerging profile of the pandemic.

The entropic maxima H_m are utilized to evaluate the Emergence Function λ for each stage of the COVID in the Western Pacific region shown as the Emergence profile in the boxed inset in Figure 4.6(b).

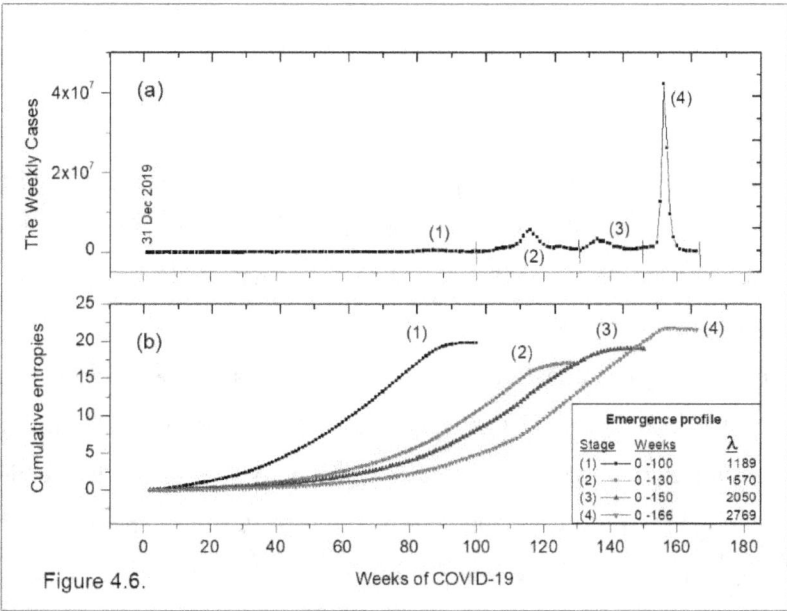

Figure 4.6.

Figure 4.6. Cumulative entropies of COVID-19 cases in W. Pacific. (a) The weekly cases for the W. Pacific region are shown to identify the four (4) distinct, slowly emerging, descriptive stages of the COVID cases over the entire 166 week. (b) Cumulative entropies are plotted for these four stages of the cumulative cases by constructing and utilizing the probability distribution functions for each stage. Note that the cumulative total for the 0-100 weeks was 10.5m; for 0-130 weeks it rose to 63.5m, the 0-150 weeks had 95.3m cases. The 4th stage (with 16 weeks added to the 3rd stage) had 201m cases. **Inset:** shows the Emergence Functional profile $\lambda = \zeta_0 * H_m$ for the set of 4 stages of COVID-19 in the W. Pacific region.

The boxed inset of Figure 4.6(b) shows the data for the Emergence Function λ for each successive stage. The data is further tabulated in Table 4.3 along with the values of ζ_0. The course of spreading of the Coronavirus in W Pacific region had its unique features.

TABLE 4.3. W. PACIFIC

CUMULATIVE WEEKS	ζ_0	H_m	$\lambda = \zeta_0 * H_m$
0 – 100	65.43	18.17	1189
0 - 130	95.15	16.5	1570
0 – 150	107.9	19.0	2050
0 –166	125.3	22.1	2769

- It was the first region to witness the emergence of the Coronavirus. Thereafter, COVID-19 spread in all directions and regions. The emergence is measurable in the spatial (country-wise) and temporal (days and week) dimensions and reported by the WHO.

- The most significant feature is the relatively smaller number of cases in the 0-100 weeks ~10.5m. The rest of the regions of the world had ~250m cases.

- The number of cases during the four distinct stages of the pandemic's evolution in W Pacific gradually increased in the temporal space measured through ζ_0, from 65.4 weeks for the first stage to 125.3 weeks for the fourth stage. The four stages' respective ζ_0 are tabulated in Table 4.3.

- The entropic graphs shown in the earlier Figure 4.3(c) display four significant stages, each demonstrating the extension of the four distinct stages. All four graphs show similar growth rates of the population of cumulative cases with small variations.

- The entropic graphs H_m versus ζ in Figure 4.6(b) present the similar growth character of the cases, measure through H_m. The maximum entropy H_m~18±2 for the four graphs.

- In contrast with the similarities of the emerging COVID stages of the W Pacific, the numbers of the region's cases versus the global ones present a different scenario. Especially, when the number of W Pacific cases doubled from 95m to 201m during the weeks 154±4, the rest of the world reciprocated with lesser intensity; there was waves of ~10m cases each in Europe and Americas and ~2m in S E Asia.

- Once initiated (near the end of 2019) and managed to spread, COVID-19 pandemic had its own dynamics. COVID-19 emerged and spread like a dynamical system with constantly varying fractal characteristics. The fractal nature of the dynamical systems is based on their sensitivity to initial conditions and the constraints of the environment. It is being demonstrated here that the unique and specific-to-the-region characteristics of the pandemic are measured through their fractal/entropic, and the cross entropic parametric projections.
- The second aspect of the fractal nature of the pandemic endows it the scale invariance. Therefore, the waves of cases or the stages of evolution of the pandemic with widely different ranges of numbers of cases can be consistently compared and analyzed.

4.7. The Emergence Functions of COVID-19 in S E Asia, E Mediterranean & Africa. The three sets of graphs of the cumulative entropies as a function of ζ-the weeks of COVID-19 in the three WHO-designated region are shown in Figure 4.7(a)-(c). Utilizing the cumulative cases' profiles (Figure 4.2), the probability distribution functions for each stage of the pandemic, in the three respective regions, were constructed. These are based on the stages of the evolving pandemic, shown in Figure 4.1(e)-(g). The entropic graphs for each stage are then evaluated and have already been included in Figure 4.3(e), (f) and (g) for the three regions S E Asia, E Mediterranean & Africa.

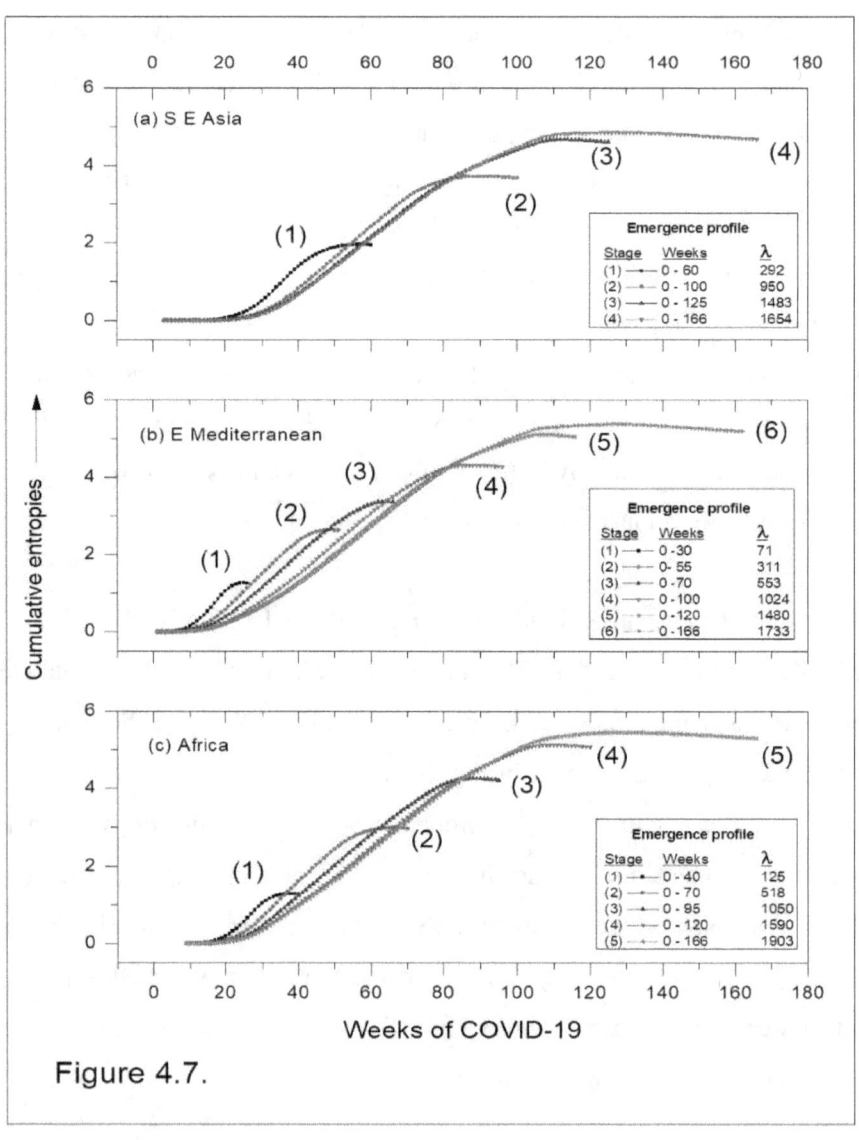

Figure 4.7. The cumulative entropies of the cumulative cases in S E Asia, E Mediterranean, and Africa. (a) Entropic graphs are plotted against the weeks of COVID-19 for the four stages of the probability distributions functions for the cumulative cases in S E Asia. (b) For the E Mediterranean region, the entropic graphs of the six (6) emergent stages of the Coronavirus cases are plotted. (c) African region displayed five (5) consecutive stages. **Insets:** Emergence profiles of the three set are tabulated as **boxed insets** in the respective figures.

These reveal the following aspects emerging from the pandemic data and the information-theoretic analysis.

- The three regions collectively contributed ~12.3% of the total cases. Out of the 93.5m cases in 166 weeks of the pandemic, S E Asia had 60.8m, 23.2m cases in E Mediterranean and 9.5m in the African region.
- The stage-wise breakup of the cumulative cases is presented in Table 4.4. The first stage, or the wave of cases took 60 weeks to generate the cumulative population of 13.4m.
- The first wave in E Mediterranean took 30 weeks to yield 1.5m cases and the corresponding one in Africa took 40 weeks to have 1.2m COVID-19 cases.
- The first wave of the cases or the stage of the evolving pandemic, indicates the efficiency of mechanisms of the spread of the virus that impacts the total outcomes.
- The peak around 110 weeks is present in all three sets of the entropic dimensional spectra. Its presence is the evidence of the mutation capabilities of the virus, as will be discussed in Chapter Eight.
- The emergence profiles of the three regions are included in Figures 4.7(a)-(c) as the boxed insets. The numerical value of the emergence function λ is calculated for each cumulative stage. These values, along with the respective ζ_0 are included in Table 4.4.
- The approximate similarity of the emergence of COVID-19 in the three regions, can be gauged from the λ values for the 166 weeks. These are 388.3 for SE Asia, 351.5 for E Mediterranean and 383.5 for Africa. The corresponding ζ_0 are 69.6 weeks, 65.4 weeks, and 70.24 weeks.

TABLE 4.4.

WEEKS/CASES	CUMULATIVE	ζ_0	H_m	$\lambda = \zeta_0 * H_m$
(A) S E ASIA				
(1) 0 – 60/13.4M		36.7	7.95	292
(2) 0 – 100/44.5M		56.1	16.94	950.3
(3) 0 – 125/58M		66.64	22.26	1483.4
(4) 0 –166/60.8M		69.6	23.97	1654.4
(B) E MEDITERRANEAN				
(1) 0 – 30/1.5M		16.6	4.29	71.2
(2) 0 – 55/5.3M		29.92	10.4	311.2
(3) 0 – 70/9.15M		37.78	14.26	553
(4) 0 –100/16.8M		52.0	19.7	1024.4
(5) 0 – 120/21.7M		61.25	24.16	1480
(6) 0 – 166/23.2M		65.4	26.5	1733
(C) AFRICA				
(1) 0 – 40/1.2M		26.6	4.7	125
(2) 0 – 70/3.3M		41.1	12.6	518
(3) 0 –95/6.16M		54.7	19.2	1050
(4) 0 – 120/8.7M		65.34	24.34	1590.4
(5) 0 – 166/9.5M		70.24	27.1	1903

The entropic and Logistic analyses of the pandemic conducted above suggest that the three regions faced similar emerging profiles of COVID-19. The differences in actual numbers of the cumulative cases can be related to the differences of the population densities of the three regions, the spectrum of the social connectivity that include the multiple transportation channels.

4.8. Summary of COVID-19 in the six(6) WHO-Regions. The pandemic of the magnitude of COVID-19 with three fourths of a billion cases and still counting, that raged for more than three years in all regions of the globe, was categorized with the Emergence Function(s) in this Chapter. Starting from a lone case of the virus, in a fish market, it reached all parts of the globe. Mutating, adapting to different social conditions and moving on; using all means of transport, by land, through the sea, and the air. The last form of human transport provided the fastest means of the virus transmission, human-to-human.

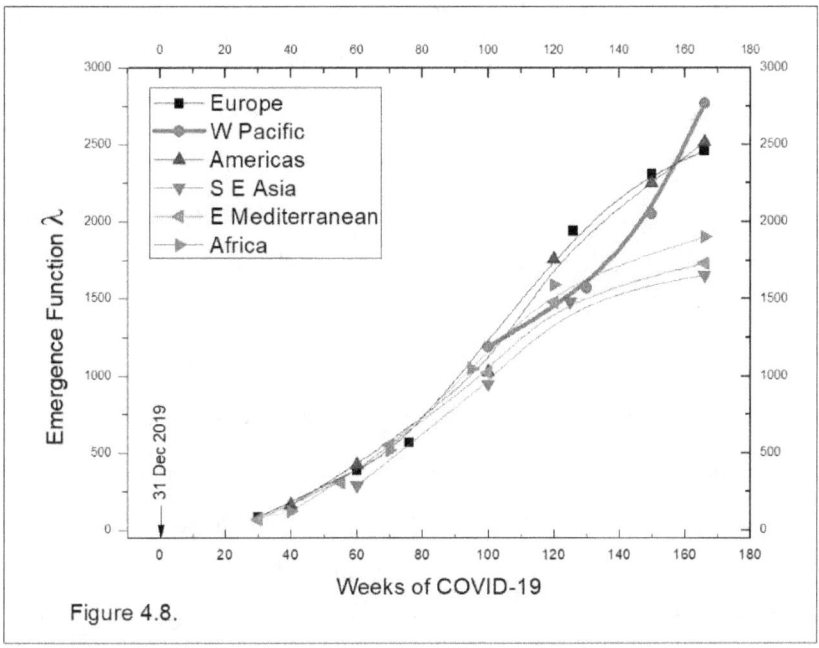

Figure 4.8.

Figure 4.8. The Emergence Functions λ are plotted for each region and the global pandemic as a function of the incremental weekly steps of COVID-19. The function is calculated from the entropic profiles for the multi-stage analyses of the six regional and the global data of the cumulative cases. The discontinuity in the 0-100 weeks of the W Pacific is highlighted.

For inclusion in Figure 4.8, the Emergence Functions were evaluated for the cumulative stages of the pandemic in the six(6) WHO-designated regions and for the global cumulative population of the infected cases. These are plotted as a function of the weeks of each respective stage. The six regions' λs and the global one, are shown in Figure 4.8. Figure 4.8 provides a direct comparison with Figure 4.2 of the cumulative cases that generated the data for Figure 4.8; with all the graphs contained in the same figure. The noteworthy features of Figures 4.8 and the Graphical Summary are:

1. The similarities of the λ versus the weeks of COVID-19 graphs for the

five regions of Europe, Americas, S E Asia, E Mediterranean, and Africa is remarkable given the different magnitudes of the number of the cases (273m European cases versus the 9.5m African cases, for example), and the diversity of the habitat (living conditions, different continents, weathers etc.) and the nature of the social and financial disparities.

2. The W Pacific region yielded comparatively less information during the first 100 weeks with ~10.5 million cases in the global total of ~262 million. However, it has similar λ for its 100 weeks, as shown in Figure 4.8 with bold symbols.

3. W Pacific region's emergence function plotted for the four stages 0-100 weeks, 0-130 weeks, 0-150 weeks, and 0-166 weeks show a sharp upward increasing trend. The graph could be interpreted as part of the ongoing pandemic that may still have to reach its holding capacity of the cases (?).

4. The graphs for W Pacific λ versus the number of weeks of COVID-19 can hardly be fitted with the Logistic or the Sigmoidal curves for not smoothly reaching the maximum or the holding capacity. The first λ is for the duration of almost two years i.e., from 0 to 100 weeks. The set of the four λ values, indicates the sharp increasing trend during the last 66 weeks. The region presents a paradox when we compare the entropic and temporal characteristics of the pandemic in other regions.

> **Graphical Summary.** The emergence profile of the pandemic with specific reference to each stage of the emerging cases of COVID-19, in all regions and the global total, with the Emergence Function λ represented as the bar graphs drawn for approximately similar temporal separations.

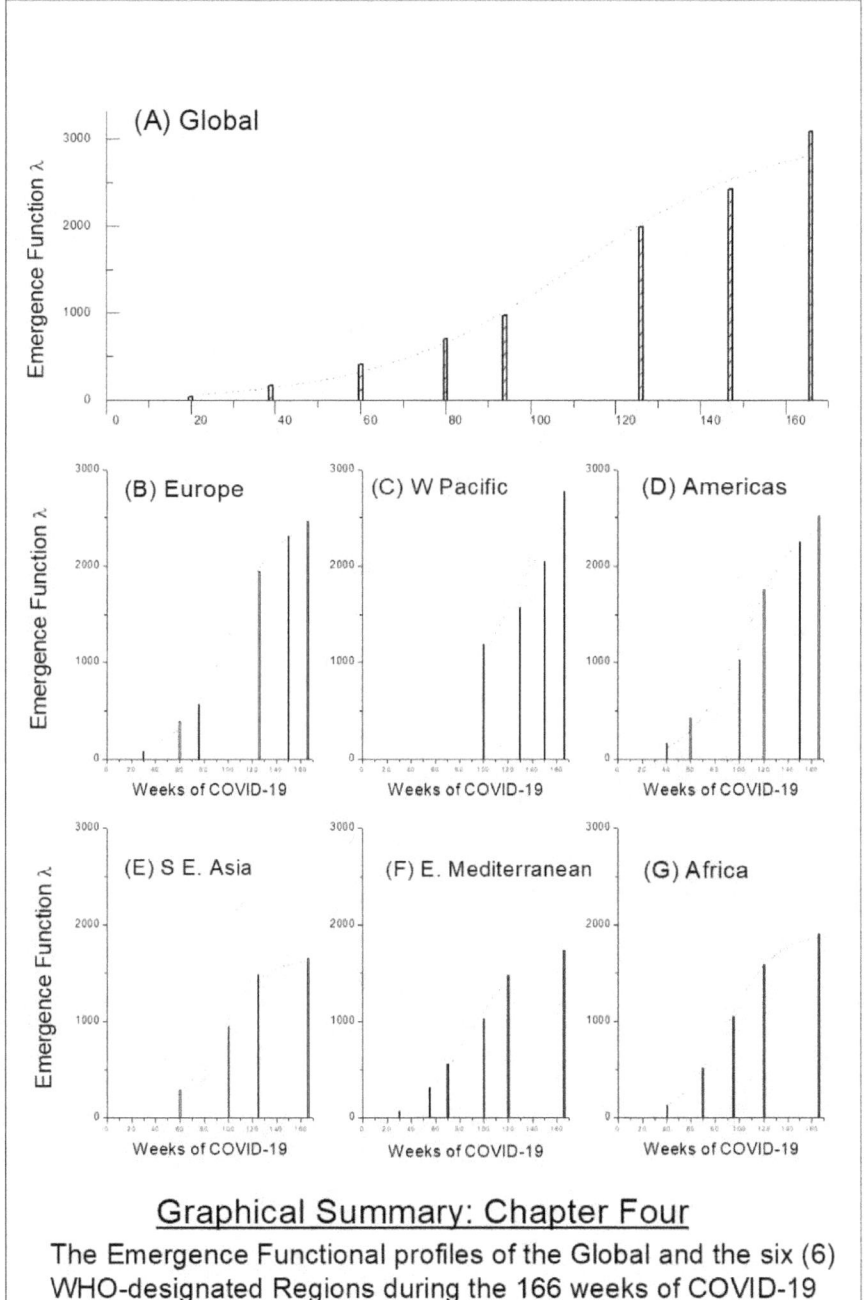

Graphical Summary: Chapter Four

The Emergence Functional profiles of the Global and the six (6) WHO-designated Regions during the 166 weeks of COVID-19

CHAPTER 5

CROSS EMERGENCE OF COVID-19

5.1. Cross Entropic Profiles of the Emerging Pandemic. The initiation, multiplication of the virus-infected cases, and the spread of a pandemic can be described by the twin, emergent and complementary, dynamical mechanisms of Emergence and Cross Emergence, introduced in the first two chapter and illustrated in Chapters 3 and 4. The information-theoretic emergence model developed in Chapter One and applied in Chapters Three and Four for investigating the profiles and modes of the spread of Coronavirus in 2002-4 and 2019-, utilized the Emergence functions λs evaluated through the respective inflexion points ζ_0s obtained from the graphs of the Logistic growth and the associated H_m from the respective Shannon entropic graphs. The analysis was based on evaluating the probability distribution functions $p(\zeta)$ constructed for the cumulative global cases, and similarly, for the WHO designated regions, for each stage of the pandemic. The set of λs utilized the number of cases of the emerging waves of the infected cases. The data was obtained from the numbers of the emerging COVID-19 cases on the WHO web site [70,71]. The evaluated probability distributions contain the basic, emerging information of the cases during the successive stages of the pandemic. The accumulating cases for a certain period represented by an appropriate probability distribution function of the cumulative cases, were constructed by the procedure described in Chapter Four and the earlier ones. This procedure for evaluating the $p(\zeta)$ for the specific time or the stage of the data collection, plays a significant role when the multiple regions simultaneously generate such data that lead to the consequent probability distributions. Multiple WHO regions generating their region-specific Shannon entropy and the associated Emergence functions for the specific stage of the pandemic, were evaluated, tabulated, and discussed in the previous chapter.

The technique of evaluating the Cross Emergence to be employed here for COVID-19, was developed in Chapter Two, and applied to the 1st episode of the Coronavirus in 2002-4, in Chapter Three, based on the cross entropic profiles of the pair of probability distribution functions $p(\zeta)$ and $q(\zeta)$, obtained from the data of the accumulated cases in the two different regions. The analysis implies that the condition of the temporal equivalence must be satisfied i.e., the data of the COVID-19 cases must have been acquired during the same span of time. For the intra-regional Cross-Emergent comparison, the associated Cross entropies $H(p|q)$ are employed. Each set of the two comparative probability distributions $p(\zeta)$ and $q(\zeta)$ generates the set of two consequent cross entropies $H(p|q)$ and $H(q|p)$. These were formulated, employed, and described in Chapter 2, as Eq.(2.1) and (2.2). In the context of COVID-19, the cross entropies serve as measures of the divergence between the two probability distributions $p(\zeta)$ and $q(\zeta)$ and as the quantitative analytical tools for understanding the information-manipulative role played by the circumstantial environments of the different regions. In addition, by using the cross entropies, the diverging trends of the emerging profiles of the pandemic are quantified through the probability distribution pairs that are constructed for the respective regions' cases.

The Cross Emergence Function (CEF), defined in Chapter Two, through Eq.(2.3) is constructed from the cross entropies generated by the pair of the probability distributions $p(\zeta)$ and $q(\zeta)$ and the associated inflexion points $\zeta_0(p)$ and $\zeta_0(q)$. It was shown that the Cross Emergence endows an emerging, information-generating dynamical system for example, COVID-19 in our case, the spatial and temporal characteristics through the cross entropies of the two associated probability distributions. The Cross Emergence functions $\mathcal{E}(p|q)$ and $\mathcal{E}(q|p)$ were defined by treating the distributions $p(\zeta)$ and $q(\zeta)$, alternatively, as the true and the predicted distributions with the appropriate inflexion points $\zeta_0(p)$ and $\zeta_0(q)$ generated independently, in Eq. (2.3) by the two distributions were formulated as

$$\mathcal{E}(p|q) = \zeta_0(p) * H(p|q) \text{ and } \mathcal{E}(q|p) = \zeta_0(q) * H(q|p).$$

The Index of Divergence (IoD) was also defined as the absolute value of the difference of the two Cross Emergence Functions for the set of the twin distributions. It was defined through Eq. (2.4) in Chapter Two as $\Delta\mathcal{E} = |\mathcal{E}(p|q) - \mathcal{E}(q|p)|$. The Index of Divergence $\Delta\mathcal{E}$ will be used extensively, in Chapter Seven to characterize the emerging profiles of the variants of SARS-CoV-2. The IoD will also be shown as the relevant function for describing the emergent, dynamical profile of the mutating virus during the spread of the pandemic. It will be employed extensively when evaluating the cross emergence of the SARS variants in the USA, in Chapter Seven.

5.2. The Regional Cross-Emergence of the 166-weeks of COVID-19. Figures 5.1 to 5.4 present the sets of cumulative and cross entropies for the combinations of the six-regional emergences of the pandemic relative to the global COVID-19. The six regions contributing to the global spread of the coronavirus are analyzed by the comparative analysis that employs the Cross Entropy $H(p|q)$, whereas Cross entropy is the sum of the nominal entropy and the KL-divergence (Eq.(2.1)). The Figures 5.1 to 5.4 show the cross entropies of the WHO regions calculated relative to the global normalized, cumulative probability function $g(\zeta)$. The probability distribution functions are evaluated for Europe designated as $e(\zeta)$, Western Pacific as $p(\zeta)$, the Americas as $a(\zeta)$, S E Asia as $s(\zeta)$, Eastern Mediterranean as $m(\zeta)$ and $f(\zeta)$ for Africa.

Figures 5.1 to 5.3 present the emerging profiles of the cumulative entropies and the cross-entropies for the six sets of the global and regional combinations: Europe-global, W Pacific-global, Americas-global. The three graphs for the cross entropies of (SE Asia-global), (E Mediterranean-global), and (Africa-global) are plotted in Figure 5.4(a) to (c).

The figures were prepared to present the essential features of the similarities and the subtle differences between cross-emerging pandemic and the growth of population of the cases in the form of the waves and the corresponding stages of the infected population within the WHO regions during the cumulative period of the 166 weeks to understand and demonstrate the emerging features of the pandemic. The in-depth, detailed study of the representative countries from the three regions Europe, the Americas, and W Pacific, will be conducted in the next Chapter Six. The results obtained in the present chapter will be useful in developing the Emergence and Cross-Emergence profiles. The next Chapter Six will present the analysis of the case studies in the chosen countries in the three of the WHO designated regions; China and Korea (W Pacific), Germany and Spain (Europe) and USA (Americas). The analysis will be focused on the spread of the pandemic during the initial 100 days and compared with the cumulative developments that followed during the subsequent 166 weeks.

5.3. Europe-Global cross-entropies. Figure 5.1 presents the sets of the twin graphs. The first set in Figure 5.1(a) has the European and the global cumulative entropies designated as $H(e) \equiv H(e(\zeta))$ and $H(g) \equiv H(g(\zeta))$, calculated for the COVID-19 for the entire duration of 166 weeks from the probability distributions $e(\zeta)$ and $g(\zeta)$ for the accumulating weekly cases. The set consisting of the two cross entropies calculated and shown in 5.1(b). The cross entropies $H(e|g)$ and $H(g|e)$ are derived from the sums of the respective cumulative entropies in Figure 5.1(a) and the mutually generated relative entropies or the KL-divergences. The resulting cross entropies are shown in 5.1(b) by using Eq. (2.2).

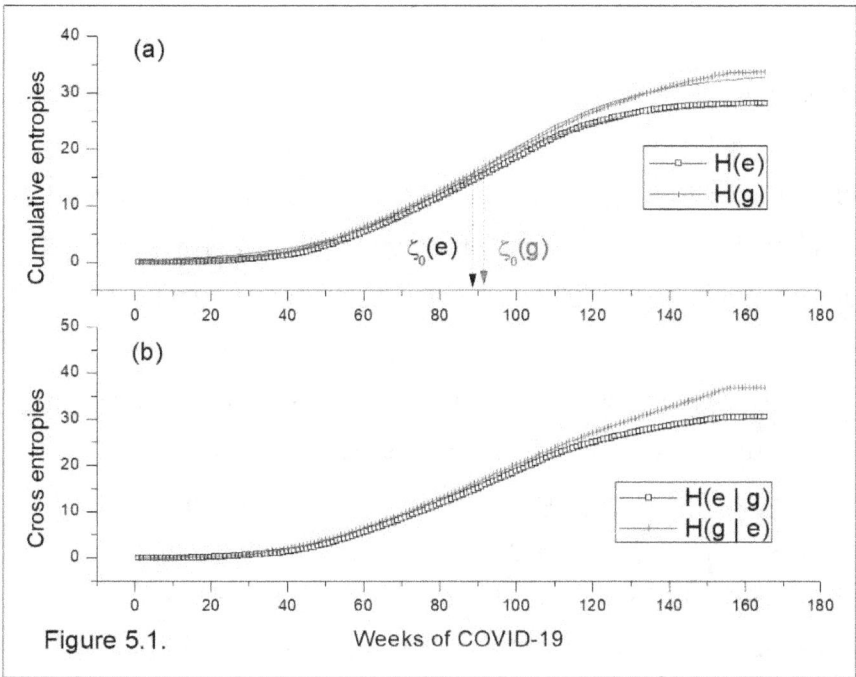

Figure 5.1. The Europe-Global cross entropies during the 166 weeks of COVID-19. (a) The set of the two Shannon entropies for European and the global cases are plotted. The two inflexion points are also shown. (b) Cross entropies $H(e|g)$ and $H(g|e)$ for the entire COVID-19 duration are shown. The close resemblance and the similarities growth in (a) and (b) demonstrate the global impact of the European probability distribution functions of the cases.

In Figure 5.1(a) and (b), the first 30 weeks generate almost identical entropies and the cross entropies. The first 30 weeks show the cross entropies are small in magnitude and approximately equal. From 40th to the 110th week, the two cumulative Shannon entropies $H(g)$ and $H(e)$ shown in 5.1(a), increase at the rates $\gamma=0.05$ for the global and a slightly faster rate~ 0.058 for the European region.

The values of the cross entropies in Figure 5.1(b) are 1.6 [cross entropic units] for $H(e|g)$ and 2.6 for H($g|e$) at the 40th week of COVID-19.

It is worth noting that the total number of cases in Europe at the end of the 40th week were 6.55 million as compared with the global total of 35.4 million cases. Similarly, at the end of the 110th week, the European region had 156.5 million cases while the global total was 395.6 million. The entropies $H(e)$ and $H(g)$ show a constant rate of increase from 1.45±0.3 on the 40th week to 23.35±0.85 on the 110th week. All units have the dimension that is measured in terms of the relative entropic and cross entropic units of the natural logarithm.

For the same period i.e., from 40-110 weeks, there was a consistently small, corresponding cross entropies increase demonstrating that the pandemic in Europe had the similar statistical and information-theoretic footprints as those of the globally spreading COVID-19 for this period. The cumulative entropic and the cross entropic profiles of this period indicate a close resemblance between the European and the global cumulative cases-driven probability distributions. The real differences between the two emerge near the later parts of the pandemic where the European cases had reached their maxima of ~270 million, but the global cases registered a massive late contribution from the W Pacific region of ~100 million cases in the period 156±4 weeks. There were no resonant increases in the remaining 5 regions. This aspect is clearly demonstrated in the later parts of the three sets of the entropic and the cross entropic graphs shown as Figure 5.1(a) and (b).

5.4. The Americas-Global cross-entropies. Figure 5.2 presents three sets of the twin graphs of Shannon entropies and cross entropies derived from the probability distributions of the cumulative cases during the 166 weeks in the Americas and the global cases. The cumulative entropies designated by $H(a) \equiv H(a(\zeta))$ and $H(g) \equiv H(g(\zeta))$ calculated from the probability distributions $a(\zeta)$ and $g(\zeta)$ of their respective cumulative weekly cases are shown in Figure 5.2(a) against the 0-166 weeks duration. The cross entropies $H(a|g)$ and $H(g|a)$ are calculated from the respective sums of the entropies in 5.1(a) and the calculated KL-divergences for the cross entropies in Figure 5.2(b).

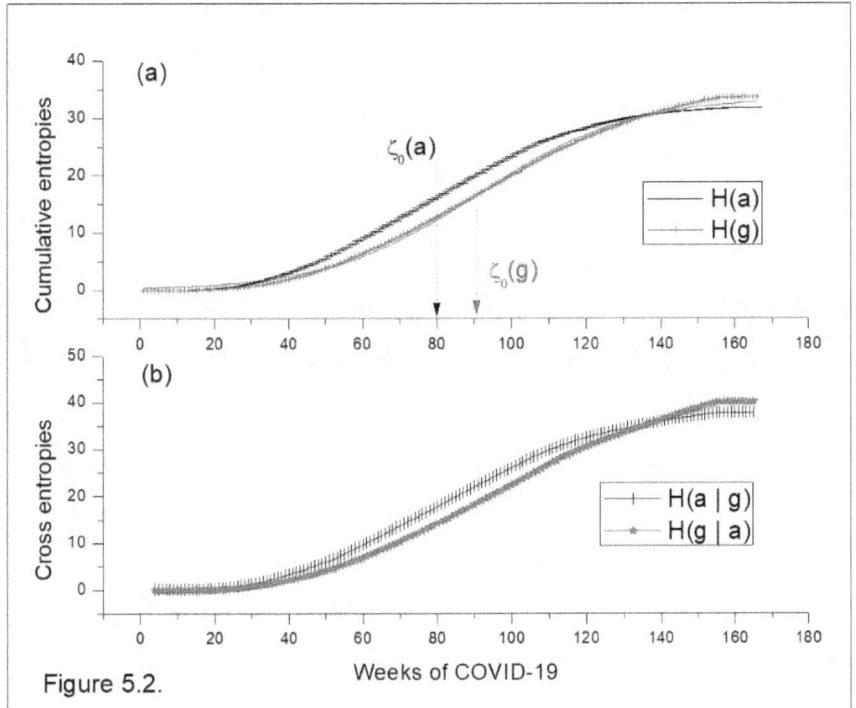

Figure 5.2.

Figure 5.2. The Americas-global cross entropic comparison. (a) Cumulative entropies of the Americas and the global probability distributions are plotted. The two crossed over around the 138th week. (b) The cross entropies $H(a|g)$ and $H(g|a)$ for the cumulative period of the 166 weeks show similarities of the growth of the cases that demonstrate the inter-global impact of the probability distribution of the COVID-19 cases in the Americas.

The entropic growth of the pandemic in 5.2(a) points to the two essential differences between the Europe-global (shown in 5.1(a)) and the Americas-global information-theoretic profiles. Firstly, the global cumulative entropy is > the European entropy during the 166 weeks of COVID-19, and secondly, the cumulative entropy for the Americas is > that of the global until the cross over around the 138th week when the entropic graph for the Americas reaches its holding capacity ~30 [entropic units]. Global entropy continues to increase to ~33. At the cross-over point occurs around $\sum_0^{138} a(\zeta)\ln(1/a(\zeta)) = \sum_0^{138} g(\zeta)\ln(1/g(\zeta))$. This is the entropy equalizing limit (EEL) of Chapter 2 with two significant differences: (1) This EEL stage has not been approached solely between the two information exchanging dynamical systems as emphasized in Chapter Two, rather it is between the regional reservoir and the global one which is built up by the collection of 6 regional reservoirs of cases, and (2) the entropic equality of the information generated by the two demonstrates that one of the regions (Americas) has reached its holding capacity in terms of the increase of the population of the cases, while the global total is still increasing. Here, it refers to the state of the accumulating cases that generate coherent probability distributions in the Americas and globally around the entropy equalizing stage. Beyond this stage, the global entropy continues to accumulate more cases as compared to that of the American region, until the 166th week.

The two cross entropies have almost similar, non-increasing pattern for the first 30 weeks with $H(a|g) \sim H(g|a) \sim 0$ and diverge for >35 weeks with the $H(a|g)$ showing a consistent pattern of $H(a|g) > H(g|a)$ up to the 138th week when the two equalize. For the most part of the 166 weeks, the two demonstrate $H(a|g) > H(g|a)$. At the cross over in the 138th week, the two cross entropies also equalize and $\Delta[H(a|g) - H(g|a)] = 0$. The cumulative entropies and cross entropies equalizing phenomena is consistently present in Figures 5.2(a) and (b).

The cross entropy of the Americas shows a higher, comparative increase from the 30th to the 140th week. The cross entropies $H(a|g)$ and

$H(g|a)$ for the 166 weeks present the broad similarities of the growth of the cases and demonstrate the inter-global impact of the emerging probability distribution of the Americas.

5.5. W Pacific-Global cross-entropies. The Western Pacific region displayed a significantly different pattern of evolution of the pandemic. The profiles of the weekly and the cumulative cases has already been shown in Chapter 4 as Figures 4.1(c), and 4.2. The four entropic stages of the COVID-19 were shown in Figure 4.3(c). The cumulative entropic graphs for the whole duration 0-166 weeks were again plotted in Figure 4.6(b) to evaluate the Emergence function λ. Here, in Figure 5.3(a) and (b), the entropies and cross entropies are replotted for W Pacific region along with those of the Global probability distribution of the cases. The two graphs for the Shannon cumulative entropies diverge significantly after ~30 weeks with the global entropy $H(g) > H(p)$ for the remaining period.

The resulting cross entropies $H(g|p)$ and $H(p|g)$ start to diverge after the 30th week of the pandemic. The maximum $\Delta[H(g|p) - H(p|g)]$~32.6 units at the 110th week. The cross entropies reach their maximum at the 166th week with $H(g|p)$~75.5 cross entropic units and the corresponding $H(p|g)$~50.5. For the sake of comparison, in Figure 5.2(b), the global-Europe cross entropy $H(g|e)$~37 was larger than the corresponding Europe-global cross entropy $H(e|g)$~30. Figure 5.2 for the case of the Americas-global cross entropy $H(a|g)$ and the global-Americas $H(g|a)$ demonstrate a different profile of smaller divergence and the crossover as explained in section 5.1. The emergence profile of the COVID-19 in the Western Pacific region has unique features. The Emergence Function λ calculated for the region displays no significant growth of the cases during the 0-100 weeks as was pointed out in Chapter 4 in Figure 4.8 and the Graphical Summary. The cross entropic analysis presents the same profile of massive divergence with the global profile, as compared with the equivalent divergences of the regions of Europe and the Americas with the global one.

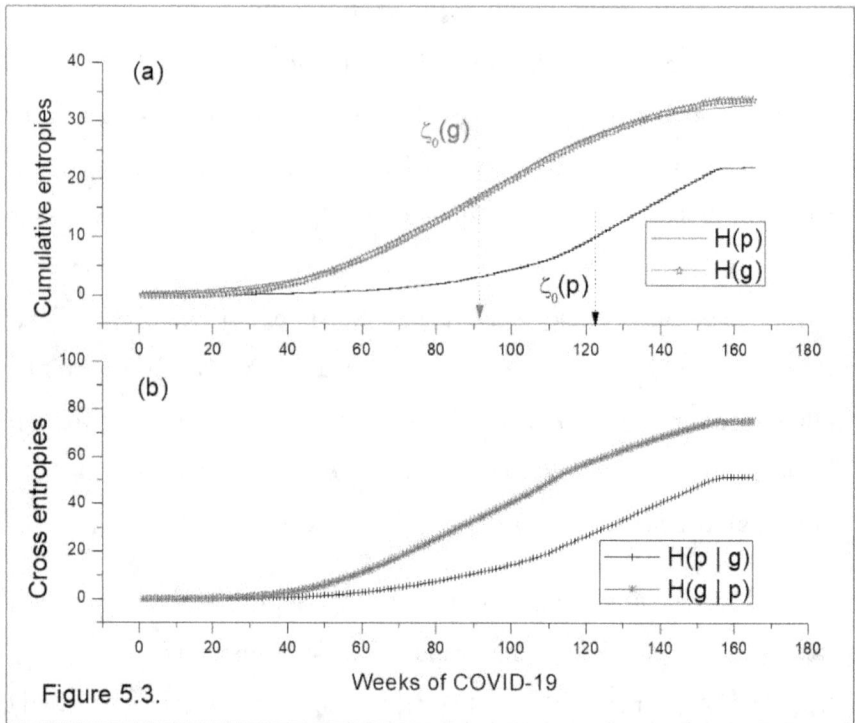

Figure 5.3.

Figure 5.3. W Pacific-Global cross entropic analysis. (a) Cumulative entropies calculated from the Western Pacific and the global probability distributions are plotted as a function of ζ-the weeks of COVID-19. The two diverge around ζ-30 weeks and continue to diverge. (b) The cross entropies $H(p|g)$ and $H(g|p)$ for the 166 weeks demonstrate the similar diverging pattern of growth. That reaches maximum divergence within the 100-130 weeks and stays divergent until the 166th week.

Western Pacific region's unique position among the WHO-designated regions will be further investigated in the next Chapter Six where a detailed analysis of the cases during the first 100 days will be contrasted with those during the later period and the cumulative effects of the 166 weeks' pandemic will be investigated.

5.6. Cross-entropies of the Global and the S E Asia, E Mediterranean, and Arican regions.

Figure 5.4 combines the cross entropic data of the three regions with respect to the global. Figure 5.4(a) shows the cross entropies for the S E Asia-Global, (b) for the E Mediterranean-Global and (c) the Africa-Global. Cross entropies are calculated by treating each pair's probability distributions, alternatively as true and predicted, as explained in section 5.1.

Figure 5.4 (a) to (c) are constructed by using Eq. 2.2 of Chapter Two. It showed that (Nominal cumulative entropy + KL divergence) → Cross entropy. Figure 5.4(a) shows the Global versus S E Asian cross entropies $H(s|g)$ and $H(g|s)$ calculated from the probability distributions $s \equiv s(\zeta)$ and $g \equiv g(\zeta)$. Here, the sum of nominal Shannon entropies of the individual distribution and the relative entropies with respect to the global distribution $D(s \parallel g)$ and $D(g \parallel s)$, are used to plot the graphs for the cross entropies $H(s|g)$ and $H(g|s)$.

Figure 5.4(b) plots the E Mediterranean versus the Global cross entropies $H(m|g)$ and $H(g|m)$. These are calculated from their respective distributions of the COVID-19 cases $m \equiv m(\zeta)$, $g \equiv g(\zeta)$ and the respective relative entropies $D(m \parallel g)$ and $D(g \parallel m)$.

Similarly, the African regions' cross entropic profiles are calculated from the respective probability distributions $f \equiv f(\zeta)$, $g \equiv g(\zeta)$. The pattern of the relationship with the Global one is like the other two regions shown in Figures 5.4(a) and (b).

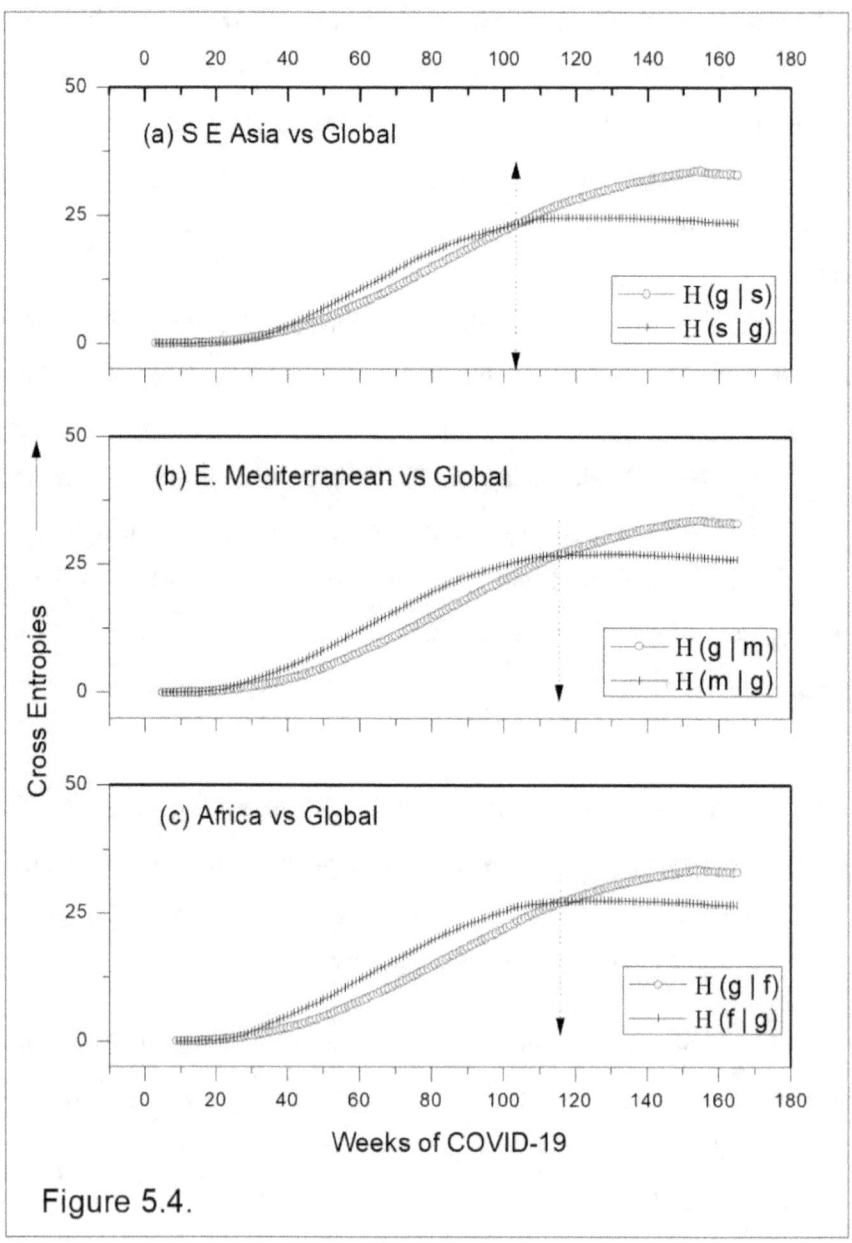

Figure 5.4. The three sets of the twin Cross entropies are plotted against weeks of COVID-19. (a) For S E Asia vs global $(H(s|g), H(g|s))$, (b) E Mediterranean vs global $(H(m|g), H(g|m))$, and (c) the African region vs

global $(H(f|g), H(g|f))$. The cross entropies are calculated with the probability distribution for the 3 regions against the global distribution function $g(\zeta)$.

There are similarities in the pattern and profiles of the emerging cross entropies for the three regions' probability distributions versus the global distribution. The basic reason for consolidating the three sets of cross entropies for the three distinct regions of the world in one figure are:

- The three sets of $(H(s|g), H(g|s))$, $(H(m|g), H(g|m))$, and $(H(f|g), H(g|f))$ demonstrate similar cross entropic profiles.
- The regional cross entropies generate higher cross entropies than the global one up to the cross entropy equalizing limit or the cross-over point.
- The regional cross entropies crossover in the 104th week in the case of S E Asia in Figure 5.4(a), and for the E Mediterranean and the African regions, the cross over point is approximately ~116th week. Here, the cross entropies equalize, $\Delta [H(s|g) - H(g|s)] \sim 0$.
- There is the mutual maximum cross entropic divergence $\sim 5 \pm 0.5$ units in the three probability distributions of the cases with the global distribution.
- The sequence of figures (5.1) to (5.4) has presented the comparative emerging profiles of the pandemic in the six WHO designated regions along with the emerging profile of the global COVID-cases during the 166 weeks. The first three figures have plotted three sets of the cumulative entropies and the resulting cross entropies for the three sets of regional probability distributions $e(\zeta)$, $p(\zeta)$ and $a(\zeta)$ against the global $g(\zeta)$.

The last figure (Fig. 5.4) plots only the cross entropies demonstrating less divergent profiles.

5.7. Global and the 6 Regions' Cross Emergence Functions.
Table 5.1 has the Cross Emergence Functions (CEFs) evaluated for the data in Figures 5.1 to 5.4. The CEFs denoted as $\mathcal{E}(g|x)$ and $\mathcal{E}(x|g)$ where $x(\zeta)$ denotes the six regional distributions, are evaluated for Global and the regions, with their respective probability distribution functions $g(\zeta)$, $e(\zeta)$, $p(\zeta)$, $a(\zeta)$, $s(\zeta)$, $m(\zeta)$, and $f(\zeta)$.

The degree of the divergence of the cross entropic profiles, elucidated above, yields the data that further elaborates the divergent nature of the pandemic in different regions. It must be pointed out that the pairs of CEF for a broad range of cumulative data for the entire 166 weeks will yield gross results. To account for the individual waves of the cases over specific time periods is likely to yield a better cross emergent profile, as will be shown in the next chapter.

The Emergence Functions were plotted in Figures 4.8 of the previous chapter. These are re-plotted in the Graphical Summary, at the end of this Chapter, to indicate the comparative emergence profiles of each of the WHO regions along with the global emergence. The Cross Emergence Functions deliver the inter-related, global pandemic in the form of the comparative information generating characteristics of the constituent regions. The Emergence Functions of the six combinations of the Cross Emergence Functions are tabulated in Table 5.1, for the computed values of the twin sets for

$[\varepsilon(g|e), \varepsilon(e|g)]$, $[\varepsilon(g|p), \varepsilon(p|g)]$,

$[\varepsilon(g|a), \varepsilon(a|g)]$, $[\varepsilon(g|s), \varepsilon(s|g)]$,

$[\varepsilon(g|m), \varepsilon(m|g)]$ and $[\varepsilon(g|f), \varepsilon(f|g)]$.

TABLE 5.1.

CROSS EMERGENCE FUNCTION $\mathcal{E}(p|q)$ AND $\mathcal{E}(q|p)$ AMONGST THE GLOBAL AND THE SIX WHO-REGIONS WITH PROBABILITY DISTRIBUTION FUNCTIONS $g(\zeta), e(\zeta), p(\zeta), a(\zeta), s(\zeta), m(\zeta), f(\zeta)$

P	q	$\mathcal{E}(p\|q)$	$\mathcal{E}(q\|p)$
$g(\zeta)$	$e(\zeta)$	3382	2694
$g(\zeta)$	$p(\zeta)$	6858	6291
$g(\zeta)$	$a(\zeta)$	3693	2911
$g(\zeta)$	$s(\zeta)$	3085	1668
$g(\zeta)$	$m(\zeta)$	3085	1733
$g(\zeta)$	$f(\zeta)$	3085	1904

W Pacific region displays a unique CEF profile. Its Emergence Function $\lambda(p)$ displays the largest value after the global $\lambda(g)$; this aspect that has been discussed in detail in the last chapter.

The cumulative entropic profiles shown in Figure 4.3(e) to (g) of the three regions of S E Asia, E Mediterranean, and Africa demonstrate approximately similar emergence trends; the cross entropies of these regions with respect to the global pandemic (Figure 5.4) also show similar patterns, as discussed earlier. The cross entropies $H(s|g)$, $H(m|g)$, and $H(f|g)$ reach approximately similar maxima ~25 within ~100 weeks. All three regions have CEEL. The three regions demonstrate similar orders of magnitude 'Sink-like' emerging pattern discussed in Chapter One, with the global pandemic seems to be operating as the 'global-Sink.' This occurs even though the S E Asian region had a total of ~61 million, E Mediterranean 23.2 million and Africa had only 9.5 million COVID cases as compared with the global total that approached ~760 million in 166 weeks. The Source-Sink analogue of the model in Chapter One can be adapted for the situation of the emerging pandemic in these three regions. Here, the mutated variants seem to have caused the successive waves of the cases, as shown in Figure 4.1(e) to (g) in an analogous way these produced the respective peaks in European region and the Americas. The Emergence function λ with similar profile for the five regions, except W Pacific, for the first 100 weeks. The European and American regions produced a faster rate of increase for their Emergence functions λ as compared with the other three regions. The uniquely divergent profile of the Emergence function λ for the Western Pacific, has been discussed above and in Chapter 4.

5.8. Summary. The chapter has elucidated the twin cross entropies evaluated for any two probability distributions of the temporally coexisting emerging dynamical systems that are the byproducts or the progenies of the common ancestor, are expected to display the features that depend upon the conditions, the environmental factors, and the region-specific pandemic-control parameters. Let us summarize the cross-emergent profiles of the six regions as compared with, and referenced to, the global one.

- The cross entropies of European region and the global cases, in Figure 5.1(c) have no cross-over throughout the 166 weeks; both continue to increase at the rates that do not allow one to recede and allow the cross-over. This implies that there is no Cross Entropy Equalizing Limit (CEEL) that would indicate approach to a stable regime of the pandemic between the two information-exchanging constituents, as had been the case for the Entropy Equalizing Limit (EEL), defined in Chapter 2 between the information exchanging Sources and Sinks. Here, the European region represents one of the six sources that operate as sinks of COVID cases; whose cross entropic profile is being plotted with the cumulative sink of the global cases. No crossing of the cross entropic graphs in Figure 5.1(c) and approximately similar rates of increase of $H(e|g)$ and $H(g|e)$ as a function of ζ, implies that the gross features of the emerging pandemic in the European region truly represented the global profile in the period from the first to the 120th week. This is evident from the entropic and their cross entropic profiles.

- The region of the Americas produced the profile of a comparatively faster growth of the pandemic for the first 100 weeks as compared with the global profile and evidenced by the entropic graphs in Figure 5.2(a). The entropic cross-over or the Entropy Equalized Limit (EEL) between the two cumulative entropic graphs occurs around ~135th week. A similar cross-over occurs in Figure 5.2(c) for the cross entropies. The divergence between the two cross entropies reached maximum during the period 70-100 weeks and the magnitude is twice that for the European-global graph Figure 5.1(c).

- The Western Pacific-global entropic and cross entropic graphs of Figure (5.3) demonstrate a vastly different emergence profile of the COVID-19 pandemic. The entropic graph for the region grows slowly and so is the case for the cross entropic graph $H(p|g)$. The cross entropic divergence between $H(p|g)$ and $H(g|p)$ is more than five times that of the corresponding Americas-global cross entropies in Figure 5.1(c). Western Pacific displays a different profile for the pandemic after the first 30 weeks. It shows a pandemic under control; lesser number of cases, slower rates of growth of the cases as evidenced by the entropic profile, and a region-specific emerging pandemic that seems too distant in terms of the growth of the cases as compared with the global pandemic.

Graphical Summary. (a) The average value of the Emergence Functions of the pandemic in six WHO-regions and across the globe are of the magnitude ~2300±700. The three regions of S E Asia, E Mediterranean and Africa generate lower λs as compared with the other three regions. (b) The Cross Emergence Functions derived from the Probability distributions of the COVID-19 cases in six(6) WHO-designated Regions against the Global distributions, demonstrate more pronounced divergences against the global distribution.

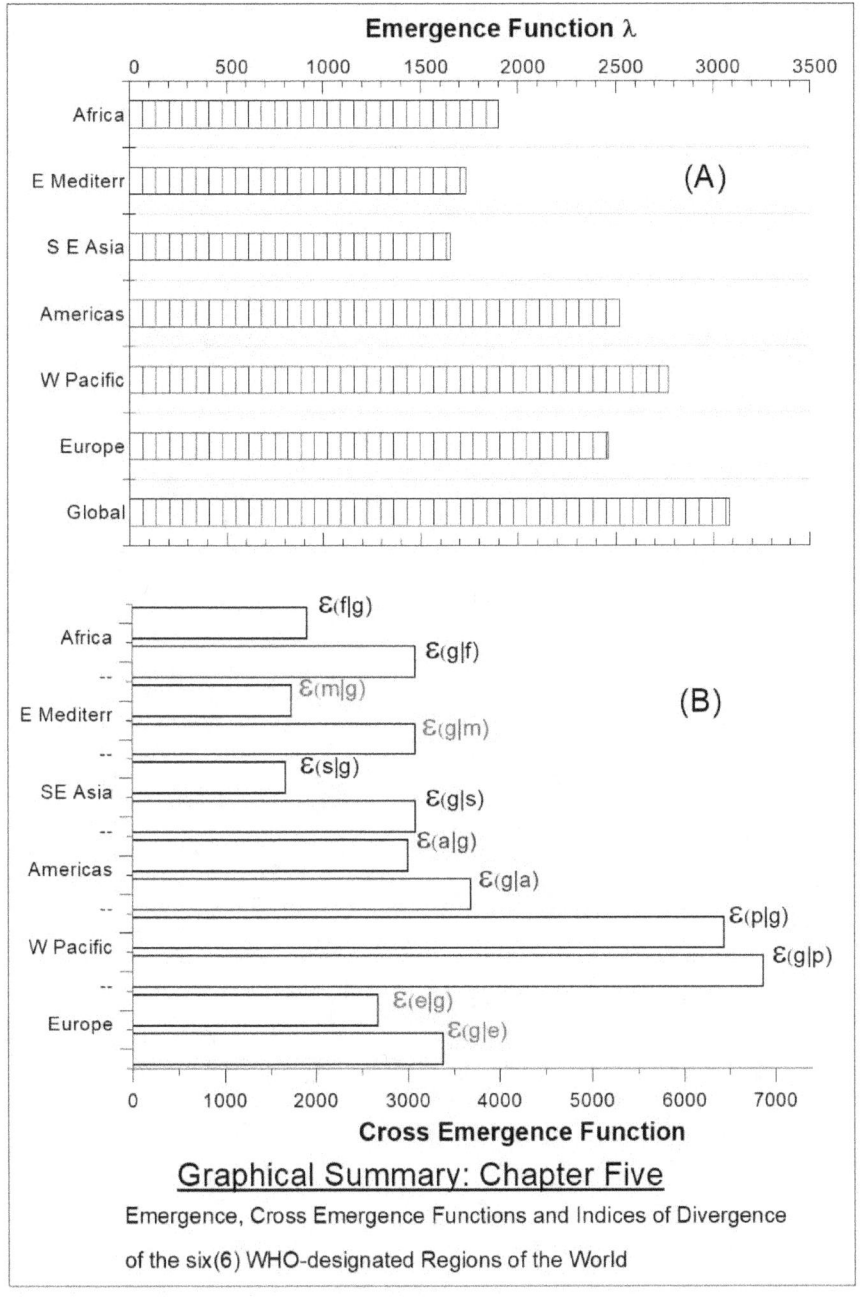

Graphical Summary: Chapter Five

Emergence, Cross Emergence Functions and Indices of Divergence of the six(6) WHO-designated Regions of the World

CHAPTER 6

THE FIRST 100 DAYS & THE 166 WEEKS OF COVID-19 IN FIVE(5) CHOSEN COUNTRIES

Case studies of China & Korea (W Pacific), Germany & Spain (Europe) and the USA (Americas)

The profiles of the emergence and spread of the COVID-19 in the chosen countries of the three WHO regions, are reviewed here by utilizing the tools developed for the application of the Source-Reservoir-Sink (SRS) model developed in the first two chapters. The model was applied to the earlier episode of the SARS-CoV-2 pandemic in Chapter 3. Similarly, the global and the regional emergence of the COVID-19 cases during the 166 weeks has been analyzed in Chapter 4, followed by the Cross Emergence of COVID-19, in Chapter 5. The present chapter deals with the first 100 days and the cumulative 166 weeks (1162 days) in the five (5) chosen countries China, Korea, Germany, Spain, and the USA of the three WHO-regions. The investigations are conducted in two distinct phases: (a) the localized Emergence and (b) Cross Emergence vis a vis the global pandemic. The information-dimensional diagnostic tools for the study of local and global pandemics are the Emergence and the Cross Emergence Functions. These have been introduced and defined in the previous chapters. Index of Divergence (IoD) $\Delta \mathcal{E}$, as indicators of distinctive divergence, will be expressed through the Cross Emergence Functions $\mathcal{E}(p|q)$ and $\mathcal{E}(q|p)$ for the probability distributions $p(\zeta)$ and $q(\zeta)$ of the Coronavirus cases.

Section (A): The information-theoretic analysis of the initiation and spread of the COVID-19 conducted during the initial 100 days in the chosen countries of the three of the WHO-regions of W Pacific, Europe and

Americas is presented. China and Korea, Germany and Spain, and the USA have been the chosen, representative countries of the regions.

Section (B): The detailed investigation of these five countries spanning the entire 166 weeks of COVID-19, are presented employing the same diagnostic tools of Emergence, Cross Emergence Functions and the IoDs $\Delta\varepsilon$.

(A) The First 100 days

6.1. The First 100 days of COVID-19 in the W Pacific Region: case study of China and Korea. The WHO's 2nd Situation report dated 22January 2020 [80] noted that 'As of 21 January 2020, a total of 314 confirmed cases have been reported for novel coronavirus (2019-nCoV) globally. Of the 314 cases reported, 309 cases were reported from China, two from Thailand, one from Japan and one from the Republic of Korea'. That is how the WHO-reported the initiation of the Coronavirus, which was later redesignated as COVID-19.

The two countries of the W Pacific region chosen for analysis in this chapter are China and Korea. These were chosen for not only being the first where the pandemic's initial spread occurred, but also for being the ones where the pandemic demonstrated a unique profile of the quick and early emergence followed by the equally robust controls by the respective governments on the spread, as was indicated, and discussed in section 4.6 of Chapter 4.

Figure 6.1 presents the pattern, the profile, and the entropic analysis of the initial stages of the pandemic in China, and Korea, and compared with the emerging global profile. In Figure 6.1(a), the cumulative cases have been plotted as a function of ζ-the days of pandemic. The logarithmic plot of the cumulative cases spans seven (7) orders of magnitude from 10^0 to $\sim 10^7$. The logarithmic graph helps to highlight the nature of transition of the initial phase from the few cases to thousands, and then lead to the millions of cases. Day 1 shows one case in Korea, ~300 in China and ~300 for the global count. Within the next 20 days, Chinese cases approach 83,000 and then increase at a much slower, controlled pace, for the next 80 days. The Chinese pandemic was controlled through various methods and means of social distancing, workplace procedures, quarantines, and similar other measures.

COVID-19 in Korea was characterized with open information, public participation and wide-spread testing that formed the basis of the strategy by the Korean government from the day the first case of a 35-year-old Chinese woman was reported [80]. In Figure 6.1(a), the Korean cases' cumulative graph can be seen to rise to ~ 350 cases in 30 days. That rise was attributed to the worshippers at Shincheonji Church in Daegu which became the local Reservoir and the Source of the cases. The next 20 days saw the total increase to ~11,000. The Koreans managed to control the onset of their local pandemic and the emergence of the resulting Reservoir/Source of cases that showed nominal increase of the cases from the ~40th day onwards.

The pandemic-management plan, strictly followed by the Koreans [82] was (a) to identify the infected person(s), (b) trace all those who may have been in contact with the infected persons and then quarantine them. Thermal cameras were installed at airports and the entrance points to the suspected places followed by wide-spread testing, tracing the contacts, isolation/quarantine, and medical support to the infected. During the period 20 January to 30 April 2020, no country-wide lockdown was enforced, however precautionary measures were introduced that included measuring of the body temperatures at all public places, buildings, and shops all over the country. However, only in Daegu, public libraries, museums, churches, day care centers and courts were closed to combat the emerging pandemic.

In addition, 2-4 weeks' delay was introduced in the new semesters of schools and universities.

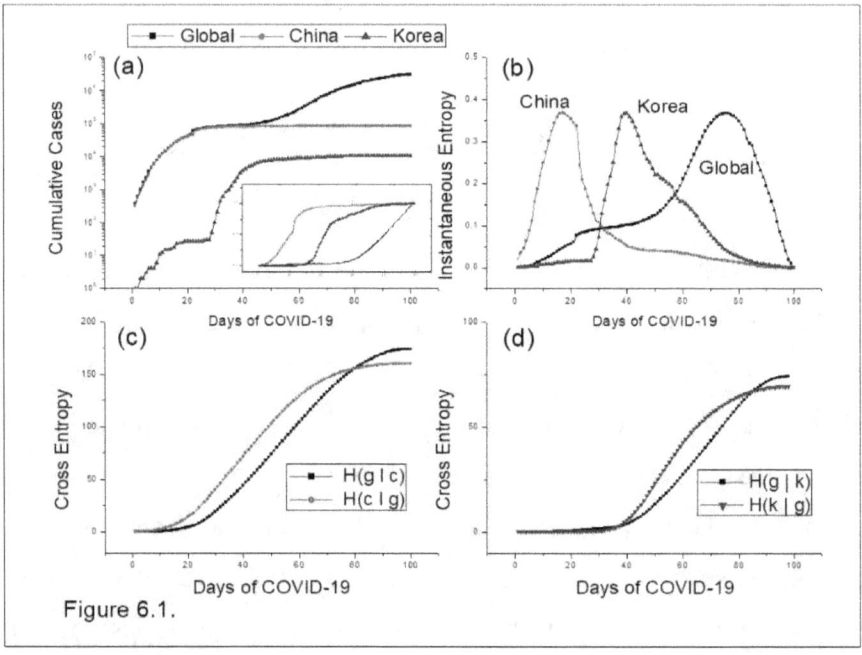

Figure 6.1.

Figure 6.1. The 1st 100 days of COVID-19 in China, Korea, and the rest of the World. (a) The emergence of the Coronavirus cases in China, and soon after in Korea, are compared in this figure with the cumulative global cases. The cumulative Chinese cases were 84373, 10,765 for Korea and ~ three million global cases in the first 100 days, starting from 21 Jan to 30 April 2020. **Boxed Incet:** shows the cumulative probability distributions. (b) The instantaneous entropic plots of [probability*ln(1/probability)] for Chinese, Korean, and the global probability distributions $c(\zeta)$, $k(\zeta)$ and $g(\zeta)$, identify periods of the maximum COVID activity for China and Korea. The two peaks $[c(\zeta) * \ln(1/c(\zeta))]$ and $[k(\zeta) * \ln(1/k(\zeta))]$ occur earlier in the sequence leading to the rising peak for the graph of the global $[g(\zeta) * \ln(1/g(\zeta))]$. (c) The two cross entropies $H(g \mid c)$ and $H(c \mid g)$ between the probability distributions $g(\zeta)$ and $c(\zeta)$ are plotted as a function of ζ- the days of pandemic. (d) The set of cross entropies between the Korean and global probability distributions $k(\zeta)$ and $g(\zeta)$ displays the comparatively lesser divergence between the respective cross entropies, as compared with those displayed in (c) between the Chinese and the global ones.

Data from the WHO web pages [80,82].

Figure 6.1(a) shows the global cases all over the world starting to rise from ~50th day onwards that eventually left the two Western Pacific countries far behind to reach ~ 3 million global cases within the next 50 days. It will be shown in the next two figures of sections 6.2 and 6.3, that the global reach of the pandemic started within the first month with the few initial cases in the regions of Europe and the Americas.

Figure 6.1(b) is prepared from the probability distributions $g(\zeta)$, $c(\zeta)$, and $k(\zeta)$ constructed for the global, Chinese, and the Korean cumulative cases, normalized to the respective maxima of the cases at the end of the 100 days, as shown in the boxed inset of 6.1(a). The figure has the three instantaneous entropy functions $[c(\zeta) * \ln(1/c(\zeta))]$, $[k(\zeta) * \ln(1/k(\zeta))]$, and $[g(\zeta) * \ln(1/g(\zeta))]$ plotted as function of the number of days of COVID-19. The Chinese and Korean graphs have their peaks at ~20th and the 40th days. Regarding the global peak at ~80th day, it must be noted that normalized probability distribution for the cumulative cases represents the growth of the population of the cases between 0 and 1, therefore, each probability distribution produces a corresponding instantaneous entropic peak. This occurs even for those distributions that have not reached their holding capacity or the maximum. The broad peak for the $[g(\zeta) * \ln(1/g(\zeta))]$ graph represents such a developing and emerging dynamic stage for the global cases. The maximum of the global instantaneous entropy does not imply that the cases had reached their peak. However, the numbers of the cumulative cases and the associated Chinese and the Korean graphs did, due to the nature of the cumulative cases' shown in Figure 6.1(a).

Figure 6.1(c) shows the cross entropic graphs $H(g \mid c)$ and $H(c \mid g)$ between the set of the two cross entropies $\Sigma\{g(\zeta) * \ln[(g(\zeta)/c(\zeta)]\}$ and $\Sigma\{c(\zeta) * \ln[(c(\zeta)/g(\zeta)]\}$. The $H(g \mid c)$ graph has two regimes: the first one for the 0-20 days with $H(g \mid c) \sim 0$, here the two probability distributions show minimum divergence, and the second regime between 20-100 days with $H(g\,|\,c)$ increasing from ~0 to ~150 cross-entropic units. The cross entropic graph of $H(c\,|\,g)$ between the Chinese and the global probability distributions has its distinctive feature of the faster increase between the 10th to the 70th day. The two graphs the cross entropic equalizing limit (CEEL) at ~80 days, it is the point where the global instantaneous entropy maximized while the Chinese instantaneous entropic graph shows its minimum in Figure 6.1(b).

The cross entropic graphs of Figure 6.1(d) between the Korean and the global probability distributions are ~0 from the first to the 40th day. The Korean instantaneous entropy has its peak at ~40th day. The two graphs start to increase and diverge after the 40th day. The maximum divergence occurs between the two cross entropic graphs during the period 40-80 days. Around the 80th day, the global instantaneous entropic graph shows a peak while the Korean had reached its minimum in Figure 6.1(b). The set of the cross entropic graphs between $k(\zeta)$ and $g(\zeta)$ displays the comparatively lesser divergence as compared with the respective global and the Chinese cross entropies displayed in 6.1(c).

6.2. The First 100 days of COVID-19 in the European Region: Case study of Germany and Spain. The few cases reported to the WHO from Germany and Spain in the first 35 days of COVID-19 can be identified in Figure 6.2(a). It has the same global graph as was in Figure 6.1(a). The German and Spanish cases increased 4 orders of magnitude within the next 35 days. On the 100th day there were 159,119 cases in Germany and 212,917 Spanish cases. The global total was ~3 million.

In Germany, the National Pandemic Plan-NPP devised by Robert-Koch Institute (RKI) was strictly followed, which advises the government on pandemics [83]. Germany is a testimony to the strict and successful implementation of their NPP by (a) containing and confining the spread of the cases and the buildup of clusters, (b) protecting its population by stopping further spread by conducting high number of tests and ensuring enough intensive care beds and the respiratory supports in the health facilities and (c) by controlling the large scale spreading of infections. Timely interventions, border closures, public awareness campaign, sufficient testing and personal protective practices seem to have helped in restricting the emergence of COVID-19 in Germany during the initial one hundred days. However, both the countries displayed similar emergence profiles shown in Figure 6.2(a).

In Spain, multiple introductions of the virus seemed to have occurred with at least two different strains of the Coronavirus [84]. These lead to the sequence of events; (a) on the 13th of March 2020, with ~4,231 cases and 120 deaths, a 'State of Alarm' for 15 days starting from 15th March was announced with all frontiers closed, (b) on the 28th of March with 64,000 cases and ~5000 deaths, the halting of all non-essential activity was imposed for further 15 days. Spain had 166,019 cases and 16,972 deaths during the next 15 days. By the 100th day of COVID-19 in Spain, the pandemic had created the reservoir of 212,917 cumulative cases. Community transmission may be responsible for large-scale outbreaks of the local transmission that occurred.

Figure 6.2(b) of the instantaneous entropy plots for the probability distributions $g(\zeta)$, $p(\zeta)$, and $s(\zeta)$ for the global, Germann and Spanish cases, demonstrate the similarities of the pandemic in Germany and Spain, however with different number of cases. Both the peaks for $[p(\zeta) * \ln(1/p(\zeta))]$ and $[s(\zeta) * \ln(1/s(\zeta))]$ at ~65th day indicate the similarities of the initiation and the mechanisms of the spread of COVID-19 in the two countries despite various levels and methods of the controls exercised.

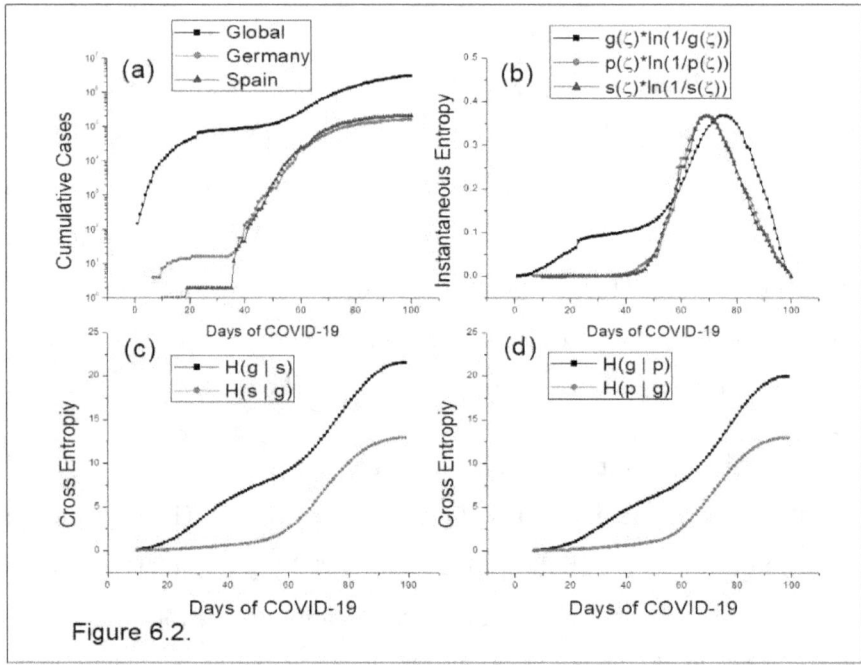

Figure 6.2.

Figure 6.2. The 1st 100 days of COVID-19 in Germany, Spain, and the rest of the World. (a) The cumulative number of the Coronavirus cases in Germany, and Spain are compared with the global cases. In Germany, the number of cases reached 159,119, 212,917 in Spain and the global total of 3 million cases in the first 100 days, from 21 Jan to 30 April 2020. (b) The instantaneous entropic plots of $[p(\zeta) * \ln(1/p(\zeta))]$ for Germany, $[s(\zeta) * \ln(1/s(\zeta))]$ for Spain, and the global $[g(\zeta) * \ln(1/g(\zeta))]$, clearly identify that the German and Spanish probability distributions produce almost identical instantaneous entropic peaks. (c) The two cross entropies $H(g \mid s)$ and $H(s \mid g)$ between the probability distributions $g(\zeta)$ and $s(\zeta)$ are plotted as a function of ζ - the days of pandemic. (d) The cross entropies between the German and the global probability distributions $p(\zeta)$ and $g(\zeta)$ display the almost similar divergence between the respective cross entropies, as compared with those displayed in (c).

Figure 6.2(c) and (d) further confirm the above conclusion. Here, the cross entropies $H(s \mid g)$ and $H(p \mid g)$ ~0 for 10-55 days and gradually rise to ~12 units during the next 40 days. The global probability distribution produces a similar order of magnitude cross entropic divergence with both the Spanish and the German distributions $H(g \mid s) > H(s \mid g)$ and $H(g \mid p) > H(p \mid g)$. This is due to the nature of the global probability distribution that is constructed from the global cases that include all six of the WHO-declared Regions, with the similar emergence profiles of the pandemic as in Germany and Spain.

6.3. The First 100 days of COVID-19 in the Americas: case study of the USA.

The US cases in Figure 6.3(a) show ~a hundred on the 40th day, increased to ~4x10^4 cases during the next 20 days. There was a total of ~1 million cases in the USA on the 100th day. It constituted one third of the global cases. USA remained the country with the highest number of cases for the remaining duration of the COVID-19.

Figure 6.3(b) presents the instantaneous entropic profiles of the global and the United States probability distributions $g(\zeta)$, and $u(\zeta)$. For the first 55 days its instantaneous entropic graph as a function of the number of the Covid days remained small i.e., $[u(\zeta) * \ln(1/u(\zeta))] \sim 0$. Its peak occurring at ~75±2 days, coincides with the global instantaneous entropic peak. Both distributions $u(\zeta)$, and $g(\zeta)$ demonstrate similar entropic profiles. This is due to the rapid increase of the cases that becomes the largest contributor resulting in the distribution for the USA being like the global distribution of the COVID-19 cases. This aspect is worth considering. The pattern, the profile, and the nature of the spread of the Coronavirus in the USA and its constituent states, seem to faithfully represent the global spread of the pandemic. It may be due to the diverse nature of the states, the topographic distribution of the population in a large, continent-like country. These may be the reasons that the pandemic in the USA simulated the global profile of the COVID-19.

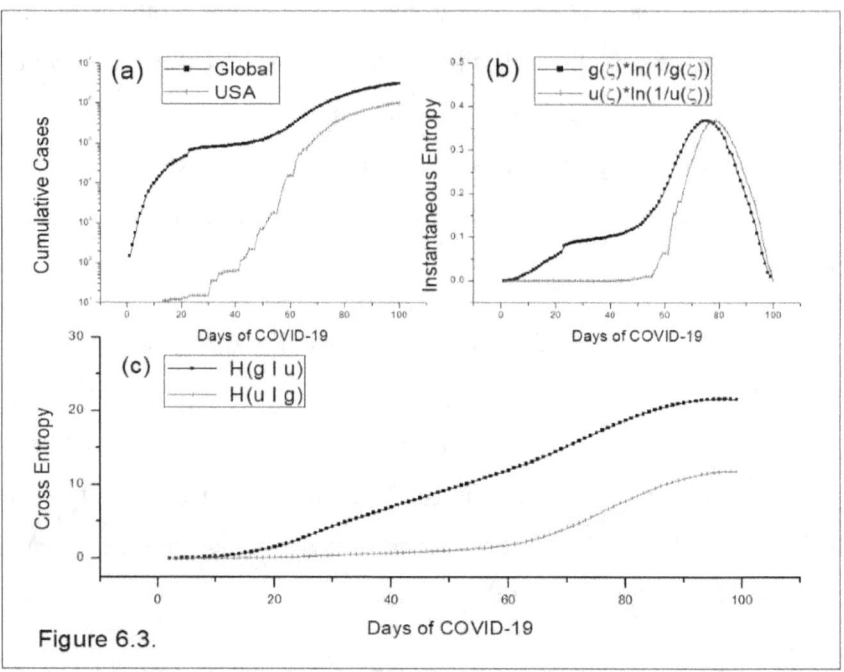

Figure 6.3. The first 100 days of COVID-19 in USA. (a) The emerging profile of the logarithmic graph of the cumulative number of Coronavirus cases in the USA are compared with the global cases. The number of cases reached ~ 1 million in USA as compared with the global total of the 3 million cases at the end of the first 100 days, from 21 Jan to 30 April 2020. (b) The US instantaneous entropy $[u(\zeta) * \ln(1/u(\zeta))]$, and the global $[g(\zeta) * \ln(1/g(\zeta))]$ demonstrate that the probability distributions $u(\zeta)$ and $g(\zeta)$ produce instantaneous entropic profiles with their peaks in their immediate neighborhood. (c) The two cross entropies $H(g \mid u)$ and $H(u \mid g)$ between the probability distributions $g(\zeta)$ and $u(\zeta)$ are plotted as a function of ζ - the days of the pandemic.

The cross entropic graph of $H(u \mid g)$ ~0 is for the period of 15-60 days. The US cases grew almost by four orders of magnitude from few to ~10^5. However, the major contributors were elsewhere. During the period 60-100 days, the cumulative cases in the USA started increasing at the rate that dominated and determined the global profile of the cases. That is also reflected in the graph of the $H(g \mid u)$. For the days 60 to 100, the rate of the increase of the cases determines the rate of increase of the twin cross entropies $H(u \mid g)$ and $H(g \mid u)$. The cross entropies of the USA-global and global-USA, show similarities for this early stage of the COVID-19 and remain so during the next stages of the pandemic, as will be discussed in the next section.

6.4. Emergence and Cross Emergence Functions for the First 100 days. For the first 100 days of COVID-19, the Emergence function λ has been evaluated to function as a diagnostic tool that characterizes the dynamical properties of the pandemic. In Tables 6.1, λ is evaluated from the inflexion point ζ_0 and the maximum entropy H_m for the five countries of this case study, by using the data presented in Figures 6.1 to 6.3.

TABLE 6.1. **FIRST 100 DAYS (0-100) OF COVID-19 EMERGENCE FUNCTION λ**

WHO REGION	Countries	Cases	ζ_0	H_m	λ
W PACIFIC	China	83,373	19.23	8.63	**166**
	Korea	10,765	45.3	9.66	**438**
EUROPE	Germany	159,119	70.2	10.21	**717**
	Spain	212,917	,,	9.92	**696**
AMERICAS	USA	1,003,974	78.05	9.96	**777**
GLOBAL		3,081,139	68.8	15.06	**1036**

It should be noted that the Emergence Function λ is a measure for the 'Emerged' dynamical systems at the end of the sequences of emergence- a function that described the entire process. Two factors that determine the Emergence function λ, are the temporal dimension (ζ_0) or the number of days of the heightened pandemic activity, and the holding capacity of the cumulative entropy (H_m) or the maximum information generated by the pandemic in a certain region or country.

The relative emergence of the five countries can be visualized by their ratios with the global emergence $\lambda(\text{global})/\lambda(\text{country})$. These are $\lambda(\text{global})/\lambda(\text{China})=6.24$,

$\lambda(\text{global})/\lambda(\text{Korea})=2.37$,

$(\text{global})/\lambda(\text{Germany})=1.44$,

$\lambda(\text{global})/\lambda(\text{Spain})=1.49$, and

$\lambda(\text{global})/\lambda(\text{USA})=1.33$.

The above table 6.1. for the Emergence Functions shows the least λ value for China. The highest value is for the USA and hence the least ratio of 1.33 for the global and the USA λs. The pandemic emerged from China, spread to Korea and then to the countries of Europe and Americas. However, the relatively smaller number of the cumulative cases in China and Korea led to smaller λs as compared with the much larger λs resulting from the higher and continuingly increasing numbers of the cases in the other three countries.

Cross Emergence Functions (CEF) $\mathcal{E}(g|x)$ and $\mathcal{E}(x|g)$ are evaluated for the pairs of distributions $(g(\zeta) \equiv g)$ and $(x(\zeta) \equiv x)$, where $g(\zeta) \equiv g$ represents the global probability distribution function and $x(\zeta) \equiv x$ is for China, Korea, Germany, Spain, and the USA, respectively. The three WHO-Regions are represented through their respective distributions $c(\zeta) \equiv c$, $k(\zeta) \equiv k$, $p(\zeta) \equiv p$, $s(\zeta) \equiv s$, and $u(\zeta) \equiv u$ designated as $(x(\zeta) \equiv x)$ in the CEF formula.

The Indices of Divergence $\Delta\mathcal{E}$ are also calculated and shown in the last column of Table 6.2.

TABLE 6.2.   **FIRST 100 DAYS OF COVID-19**
CROSS EMERGENCE FUNCTION (CEF) AND INDEX OF DIVERGENCE (IOD)

WHO REGION	Countries	Cases	ζ_0	$H(g\|x)$	$H(x\|g)$	$\varepsilon(g\|x)$	$\varepsilon(x\|g)$	$\Delta\varepsilon$
W PACIFIC	China	83,373	19.2	174	160.4	11971	3084	**8887**
	Korea	10,765	45.3	74.2	69.2	5105	3135	**1970**
EUROPE	Germany	159,119	70.2	19.96	12.96	1373	910	**463**
	Spain	212,917	70.2	21.53	12.93	1481	908	**573**
AMERICAS	USA	1,003,974	78.1	21.58	11.84	1484	924	**561**
GLOBAL		3,081,139	68.8	-	-	-	-	-

In Table 6.2, $H(g|x)$ and $H(x|g)$ are the twin cross entropies for the five combinations of the individual countries with the global distributions. The Cross Emergence Function is constructed from the points of inflexions multiplied with the maxima of the cross-entropy pair ($H(g|x)$ and $H(x|g)$) for the respective distributions. The CEF $\varepsilon(g|x)$ is evaluated by the inflexion point $\zeta_0(g)$ obtained through the relevant logistic curve fitting of the cumulative entropic graphs, and the maximum of the cross-entropy $H(g|x)$, similarly $\varepsilon(x|g)$ is determined through the relevant $\zeta_0(x)$ and $H(x|g)$.

The largest Index of Divergence $\Delta\varepsilon = 8887$ is for the global-China combination. This is due to the global-to-China CEF $\varepsilon(g|x)$ being 11971 as opposed to the China-to-global $\varepsilon(x|g) = 3084$. The global-Korean CEF $\varepsilon(g|x)$ is 5105 and the Korea-to-global $\varepsilon(x|g) = 3135$, with $\Delta\varepsilon = 1970$. Germany shows minimum IoD $\Delta\varepsilon = 463$. Spain and USA have approximately similar indices of Divergence $\Delta\mathcal{E}=566\pm6.5$.

Despite having different emergence profiles of their cases, both countries of the W Pacific demonstrate approximately similar order of magnitude CEFs $\mathcal{E}(x|g)$. The same trend is shown by the CEFs of the distributions of Germany, Spain, and the USA toward the global probability distribution. The lower values of the Indices of Divergence $\Delta\mathcal{E}$ for Germany, Spain and the USA indicate that the pandemic's emergence in these countries closely resemble the global emergence profile.

(B) The 166 weeks of COVID-19

In the previous section (A), the case studies conducted for the first 100 days of the pandemic in the chosen countries for the three regions, Western Pacific, Europe, and the Americas were presented. Here, these are being extended to the full 166 weeks duration. The cumulative global cases for the whole 166 weeks' period are included. The comparison of the relative contribution of the cumulative cases over the extended period of 166 weeks in the five countries with the global profile of the growth of the pandemic, will identify that after its emergence, the long-term evolution of the pandemic demonstrates the emergence of the self-organizing, entropic space generating, multi-component dynamical system is scale invariant and self-similar.

The data for the countries above for the COVID-19 cases have been obtained from the WHO web page https://data.who.int/dashboards/covid19/ [80].

6.5. The 166 weeks of COVID-19 in China and Korea. Figure 6.4(a) plots the data for the cumulative cases for the whole duration. The Chinese data is in the form of four (4) sharp, significant stages; the first stage is an increase from $\sim 10^2$ cases to $\sim 10^5$ in the first 3 weeks of the pandemic and stay around 10^5 for the next 105 weeks (0 to 120 weeks), the second sharp increase of an order of magnitude cases from ~0.1 million to ~0.5 million during the next 10 weeks. The third stage occurred from 120 to 152

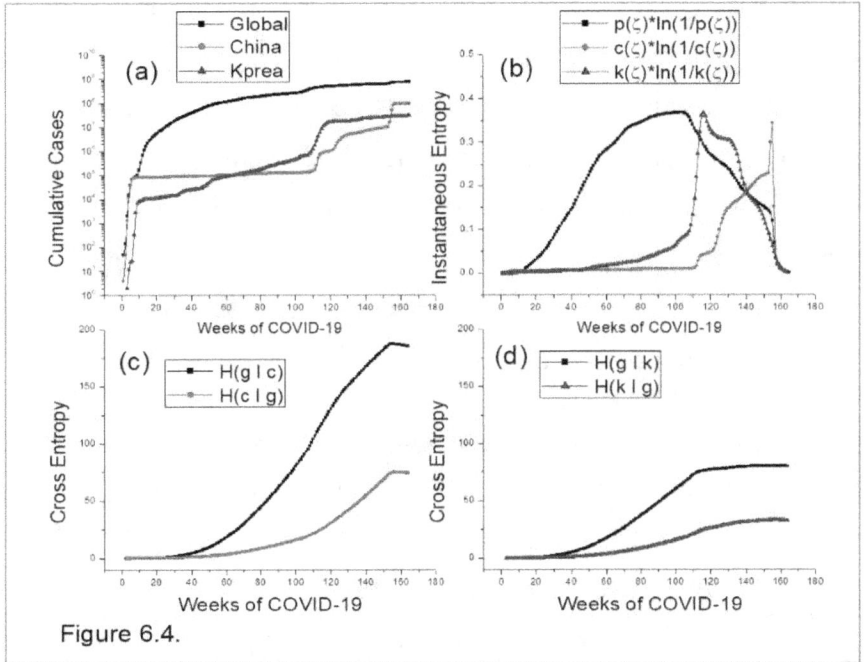

Figure 6.4.

weeks with the consequent increase to 6.7 million and the fourth occurred during weeks 150 to 160 that increased the total cases to ~100 million.

Figure 6.4. 166 weeks of COVID-19 in China, Korea. (a) The cumulative Chinese cases for the period of the 166 weeks from 21 Jan 2020 to 27 Feb 2023, were ~99 million, 30.5 million for Korea as compared with the 758.5 million global cases. (b) The instantaneous entropic plots of $[c(\zeta) * \ln(1/c(\zeta))]$ for China, and $[k(\zeta) * \ln(1/k(\zeta))]$ for the Korean COVID-19, and the global $[g(\zeta) * \ln(1/g(\zeta))]$, demonstrate the unique features of the pandemic in China and Korea versus the global one. The global peak occurs~100th week. The Korean weekly entropic graph shows a steep increase around the 110th week, goes through a sharp peak at ~120 weeks. There is the Korean period of ~40 weeks with the rise of the cases from 1.3 million to 26 million. The Chinese peak rises from the 110th to the 155th weeks from 0.146 million to 50.4 million. Another mega increase of ~90 million cases from 120th to 160th week. (c) The two cross entropies $H(g \mid c)$ and $H(c \mid g)$ are plotted as a function of ζ - the weeks of pandemic. (d) The twin cross entropies between the Korean and global distributions $k(\zeta)$ and $g(\zeta)$, display comparatively less divergence between the respective cross entropies, as compared with those displayed in (c).

The Korean cases demonstrate a sharp rising graph in the first 10 weeks from the few to ~10^4 cases. Its second stage is the continuously increasing curve for the next 100 weeks (~2 years). During the 2-year period, its cases reached ~1.3 million. The Korean cases had a third sharp increase during the period of 110-120 weeks from 1.3 to ~20 million and reached 30.5 million in the next 40 weeks. This, the last phase of sharp increase of the cases, is the characteristic Western Pacific phenomenon; it occurred in China, Korea, and the rest of the region.

The graphs for the Chinese and Korean cases shows two significant aspects; the first is the increase of ~four orders of magnitude in the number of cases (from few to 10^7) within the first half of 2020 (0-26 weeks) and the second is the relatively slower increase of ~ two orders of magnitude during the last phase.

Figure 6.4(b) plots the instantaneous entropies as a function of the weeks of COVID-19. As explained earlier in Figures 6.1(b), 6.2(b) and 6.3(b), the 'instantaneous' entropic information is best described through these plots. The $[k(\zeta) * \ln(1/k(\zeta))]$ plots demonstrate that the Korean cumulative cases generate minimal information for the 0-50 weeks. From the 50th to 110th weeks there is a gradual increase that transforms into a sharp peak within the five-week duration 110-115 weeks. This peak is followed by the gradual, stepwise decrease in the extended period 115-to-166 weeks. The Korean instantaneous entropic information graph in the third year of the pandemic starts with a sharp increase followed by a year-long stable period. This aspect is also clearly visible for the profile of the emerging cases in

Figure 5.4(a).

The trend of the Chinese graph of the instantaneous entropy $[c(\zeta) * \ln(1/c(\zeta))]$ as a function of the weeks of pandemic, during the last year is of the continuous increase from the 110th week until the final upward spike in the 155±5 weeks. It was seen in Figure 6.4(a) and discussed above.

The profile of the global instantaneous entropic graph of $[g(\zeta) * \ln(1/g(\zeta))]$ versus ζ provides the broad peak that demonstrates the global trends of the Coronavirus cases including the Korean and the Chinese contributions in the later period of 110-to-166 weeks.

In Figure 6.4(c) the cross entropic contributions of the Chinese COVID-19 against the trends of the global pandemic are plotted. The cross entropies $H(c \mid g)$ in 6.4(c) stays small in the range 0-110 weeks; this is the period of just over two years when the total Chinese cases were ~ 1million and the global cases were >400 million. The divergence of the global versus the Chinese cross entropy $H(g \mid c)$ further emphasizes this fact as it as it reaches to the maximum ~200 cross entropic units. The $H(c \mid g)$ on the other hand approaches the maximum of ~75 towards the end of the pandemic; implying that the global cases' probability distribution function $g(\zeta)$ had lesser dependence on the accumulating Chinese cases. The evolution of the Korean probability distribution function $k(\zeta)$ during the 10-110 weeks follows the trends of the pandemic elsewhere in the world, as a result, $H(k \mid g)$ the cross entropy varies from 0 to 15 in the first 100 weeks and reaches ~25 units, at the end of the 166 weeks after adding ~26 million cases in the 110-166 weeks. The global versus Korean cross entropy $H(g \mid k)$~0 to 75 units; less divergent than the corresponding $H(g \mid c)$.

6.6. The 166 weeks of COVID-19 in Germany and Spain. The emerging profiles of the first 100 days of the German and Spanish

Coronavirus cases, discussed in section 6.2, while during the 166 weeks, the pattern of growth of the cumulative cases shown in Figure 6.5(a) follow the earlier trend. The two countries had witnessed the four orders of magnitude increase in the first 100 days: from the initial few to ~10^5 cases. During the

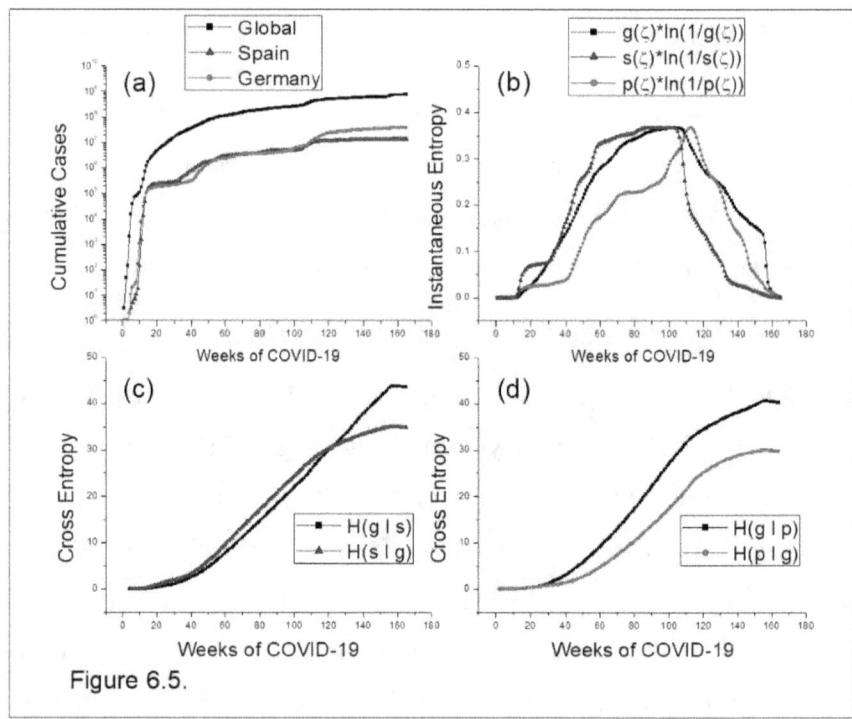

next ~150 weeks, these two countries continued the trend of the growth of the cases in Europe, as was demonstrated in Chapter 4.

Figure 6.5. The 166 weeks of COVID-19 in Germany and Spain. (a) The cumulative German and Spanish number densities of the cases for the period of the 166 weeks are ~38 million and 13.8 million, respectively. The global total was 758.5 million cases. (b) The instantaneous entropic graphs of $[p(\zeta) * \ln(1/p(\zeta))]$ for Germany, and $[s(\zeta) * \ln(1/s(\zeta))]$ for Spain, and the global $[g(\zeta) * \ln(1/g(\zeta))]$ as a function of the weeks of the pandemic. (c) The two cross entropies $H(g \mid s)$ and $H(s \mid g)$ describe similar trends in their entropic profiles with small cross entropic divergence. From the initiation of the Coronavirus to the 120th week the two graphs demonstrate almost similar profile for the growth for the pandemic cases. (d) The two cross entropies between the German and global probability distributions display the larger divergence between the respective cross entropies, as compared with those displayed in 6.5(c). There is no cross-over in 6.5(d).

In Figure 6.5(b), the Spanish instantaneous entropic graph follows the global trend while the German graph shows the large spike of the cases during the 110±5 weeks' period that increased the German total. The overall pattern of the German instantaneous entropic $[p(\zeta) * \ln(1/p(\zeta))]$ graph follows the pattern of the Spanish and the global graphs.

The Spanish and the global peaks are similar in their rising patterns; the global entropic graphs extend all the way to the end of the pandemic. The global peak occurs ~100th week. The Spanish weekly entropic graph shows a broad peak around ~80 weeks.

Figure 6.5(c) shows the gradual increase of the Spanish-global and the global-Spanish cross entropies $H(s \mid g)$ and $H(g \mid s)$ within the 0-120 weeks, further confirming the earlier entropic observation of the similarity of the growth profile of the pandemic in Spain and the global during the initial 100 days of the pandemic. The German and global probability distributions $p(\zeta)$ and $g(\zeta)$ produce continually divergent cross entropies as shown in Figure 6.5(d). The two cross entropies display larger divergence between the respective cross entropies, as compared with those displayed in Figure 6.5 (c).

6.7. The 166 weeks of COVID-19 in the USA. The first 100 days' of COVID-19 in the USA contributed ~1 million cases; and ~ five orders of magnitude increase from the first case to 10^6 cases in 100 days. Figure 6.6(a) graphically describes the 166 weeks of the cumulative cases that rise to 102 million from the first million in the first 14.5 weeks (100 days). Figure 6.6(a) shows that the USA generated the pandemic growth profile replicates the global pandemic; from initiation to the fast-spreading mode and then reaching the maximum holding capacity allowed to the pandemic by the local and the global pandemic control mechanisms.

The same conclusion is conveyed by the instantaneous entropic graphs of Figure 6.6(b). In fact, the USA graph is symmetric around the 40 to 110 weeks duration. It demonstrates the growth profile and the impact of the pandemic control measures adopted in the US. The 100+ weeks' shift in the global entropic profile is due to the Western Pacific spike in the late pandemic stages that did not have its counterpart in the regions of Europe and the Americas.

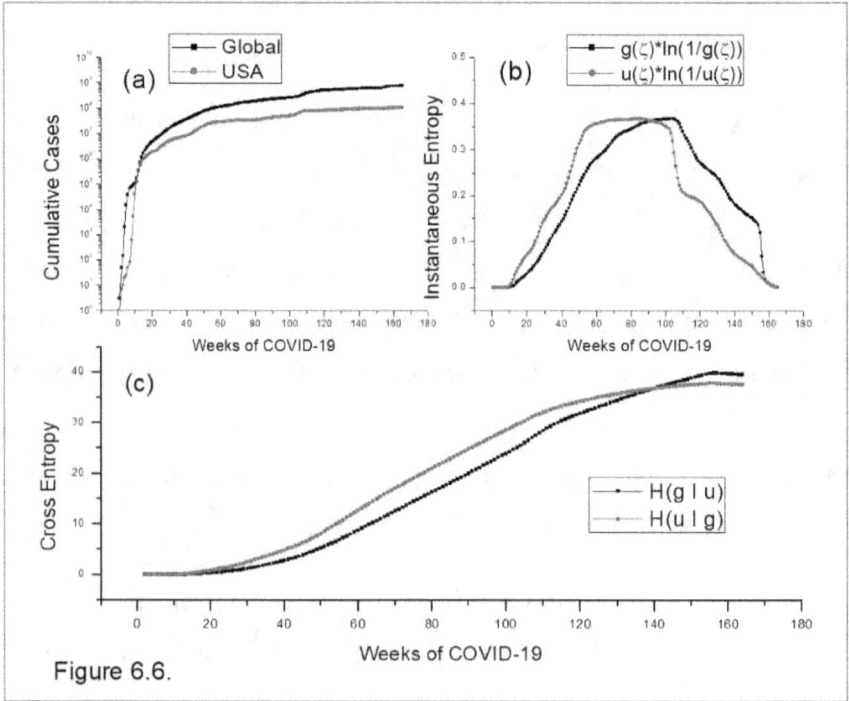

Figure 6.6.

Figure 6.6. 166 weeks of COVID-19 in USA and the Rest of the World. (a) The cumulative USA cases for the period of the 166 weeks were ~102 million as compared with the global total of 758.5 million. (b) The instantaneous entropic graphs of $[u(\zeta) * \ln(1/u(\zeta))]$ for the USA and the global $[g(\zeta) * \ln(1/g(\zeta))]$ are plotted as a function of the weeks of

COVID-19. The global peak occurs ~100th week. The USA entropic graph shows a broad peak around ~80 weeks. (c) The two cross entropies $H(g \mid u)$ and $H(u \mid g)$ between the probability distributions $g(\zeta)$ and $u(\zeta)$ describe similar trends in their entropic profiles with small cross entropic divergence. From the initiation of the Coronavirus to the 140th week the two graphs demonstrate almost similar cross entropic profiles.

The cross entropic graphs shown in Figure 6.6(c) show a gradually increasing divergence between $H(u \mid g)$ and $H(g \mid u)$ in the period between 30 and 90 weeks, which starts to reduce from the 90th week towards 140th week. The two cross entropies equalize demonstrating that the two pandemic profiles depicted through their respective distributions $g(\zeta)$ and $u(\zeta)$ generate similar cross entropies.

6.8. The stage-wise evolution of cumulative entropic plots of the 166 weeks.

The data derived from the stage-wise analysis of the cumulatively increasing cases of China, Korea from Western Pacific, Germany and Spain from Europe and the USA representing the Americas is presented in the form of the cumulative entropic graphs in Figure 6.7(a), (b), (c), (d), and (f). The graphical representation of the Emergence function λ is plotted in Figure 6.7(e).

The instantaneous and cumulative entropy plots can be used for the estimation of the inflexion point ζ_0, as has been pointed out earlier, to evaluate the Emergence Function λ for each successively increasing stage of the emerging pandemic. Figure 6.7 displays the stage-wise evolution of the cumulative entropic profiles of the five chosen countries for this case study. From each stage the corresponding inflexion point ζ_0 is obtained by fitting sigmoidal curves.

Figure 6.7(a) shows the continuity of the cumulative entropic stages with the preceding ones for the COVID-19 cases in the USA. The first stage from 0-35 weeks, had $\zeta_0 = 21.3$ weeks, 2nd stage 0-75 weeks produced $\zeta_0 =$

39.6 weeks, the 3rd stage of 0-115 weeks produced $\zeta_0 = 66.8$ weeks and the 4th stage for the period 0-140 weeks had $\zeta_0 = 73.7$ weeks. The final stage included all the earlier stages' cases from 0-166 weeks reached $\zeta_0 = 78.4$ weeks. As explained above, the probability distributions for each stage are

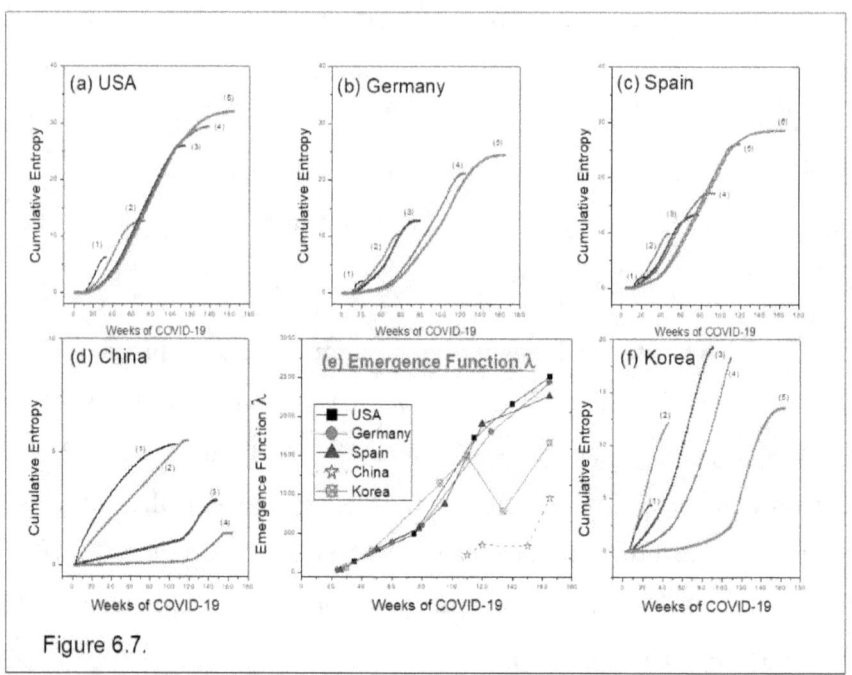

Figure 6.7.

normalized to the cumulative cases for the stage. For example, the stage-wise total cumulative cases were for stage (1) 6,515, 345. For stage (2) 33.17m, stage (3) had 79.27m, stage (4) with 94.35m and the stage (5) had 102.25m. It included all cases for the 166 weeks.

Figure 6.7. The stagewise evolutionary trails of Emergence Functions for the 166 weeks. (a) The cumulative entropies of the six (6) stages of the COVID-19 cases in USA, are plotted. (b) The same for the five (5) stages of the emerging pandemic in Germany. (c) The six (6) entropic graphs for Spain. (d) The four (4) Chinese cases' stages of evolution are shown. (e) The Emergence function λ is evaluated for USA, Germany, Spain, China, and Korea for each one of their distinct stages of the emerging pandemic versus the weeks of COVID-19. The Chinese λ displays a discontinuous graph as opposed to the other countries due to the abrupt and sharp increases shown in (d) and presented earlier in Figure 5.5. (f) The Korean entropic plots for the 5 stages are shown.

Figure 6.7(b) has the five stages of cases for Germany: the first stage (1) 0-24 weeks, $\zeta_0 = 13.7$ weeks, total cases 0.186m, stage (2) 0=60 weeks, $\zeta_0 = 37.8$ weeks, cumulative cases 2.39m, stage (3) 0-80 weeks, $\zeta_0 = 47.7$ weeks, total cases 3.73m, stage (4) 0-126 weeks, $\zeta_0 = 85.1$ weeks with 26.3m cases and (5) 0-166 weeks produced an entropic graph with $\zeta_0 \approx 100$ weeks and the total of 38.13m cumulative cases.

Figure 6.7(c) shows Spain's COVID-19 cases accumulated in six, continuously increasing stages of the cases: (1) 0-26 weeks, $\zeta_0 = 14.5$ weeks, 0.254m cases, (2) 0-50 weeks, $\zeta_0 = 31$ weeks, 1.73m cases, (3) 0-78 weeks, $\zeta_0 = 43.3$ weeks, 3.79m cases, (4) 0-95 weeks, $\zeta_0 = 51.4$ weeks, 5.0m cases, (5) 0-120 weeks, $\zeta_0 = 73.3$ weeks, 11.72m cases and (6) 0-166 weeks, $\zeta_0 = 79.4$ weeks with the total of 13.83m cases.

China displayed four distinct stages as shown in Figure 5.7(d). (1) 0-110 weeks, $\zeta_0 = 43$ weeks, 0.146m cases, (2) 0-120 weeks, $\zeta_0 = 65$ weeks, 1m cases, (3) 0-150 weeks, $\zeta_0 = 120$ weeks, 9.42m cases, and (4) 0-26 weeks, $\zeta_0 = 129.6$ weeks, 99m cases.

Figure 6.7(f) has the Korean stage-wise, graphical description of the cases. (1) 0-30 weeks, $\zeta_0 = 14$ weeks, 0.014m cases, (2) 0-47 weeks, $\zeta_0 = 24$ weeks, 0.032m cases, (3) 0-92 weeks, $\zeta_0 = 50.4$ weeks, 0.33m cases, (4) 0-110 weeks, $\zeta_0 = 81.7$ weeks, 1.29m cases and (5) 0-166 weeks, $\zeta_0 = 123.3$ weeks, 30.5m cases.

6.9. The Emergence and the Cross Emergence Functions for the total duration of 166 weeks.

In the earlier section (A) the first 100 days of COVID-19 in the five chosen countries of the three WHO designated regions, were analyzed using the entropic tools based on the respective probability distribution functions normalized to the cases accumulated during the 100 days period. In section (B) the same countries were investigated through the instantaneous entropies, cumulative and cross entropies by utilizing the renormalized probability distribution functions constructed for each successive stage of the pandemic during the 166 weeks. From the stagewise analysis of the Emergence function λ plotted in Figure 6.7, the cumulative λ for the entire duration of 166 weeks are tabulated as Table 6.3.

Table 6.3. The Emergence function λ for the cumulative data of the cases for Section (B) is tabulated for the five case studies from the sub-sections 6.4 to 6.6.

TABLE 6.3.   THE 166 WEEKS OF COVID-19 EMERGENCE FUNCTION λ

WHO REGION	Countries	Cases	ζ_0	H_m	λ
	China	99 m	139.64	7.4	**960**
W PACIFIC	Korea	30.5 m	123.3	13.53	**1665**
	Germany	38.13 m	100	24.4	**2440**
EUROPE	Spain	13.83 m	79.36	28.54	**2266**
AMERICAS	USA	102.25 m	78.4	32	**2509**
GLOBAL		758.5 m	91.53	33.67	**3082**

The pattern of the emerging profiles of λ of the pandemic during the 166 weeks in the five countries presents a qualitatively similar picture as observed during the first 100 days and tabulated in Table 6.1. During the 166 weeks, in Table 6.3, the Chinese Emergence function λ displays the lowest value, Korean λ demonstrates the next higher level for the spread of the pandemic. German, Spanish and the US pandemic displays similar emergent characteristics, with approximately similar λs. This pattern of these three countries was also observed for the initial 100 days' duration of the pandemic. The quantitative λ profiles of the five countries display similar emergence characteristics from the initiation of the pandemic (initial 100 days) to the full-blown pandemic over the 3 plus years (the 166 weeks).

Cross Emergence Function (CEF) between the global probability distribution function $(g(\zeta) \equiv g)$ and the five countries of the three regions represented through their emerging distributions $c(\zeta), k(\zeta), p(\zeta), s(\zeta)$, and $u(\zeta)$, designated as $(x(\zeta) \equiv x)$, are used to evaluate the twin sets of the CEF $\mathcal{E}(g|x)$ and $\mathcal{E}(x|g)$. The Index of Divergence $\Delta\mathcal{E}$ for each twin set of the CEFs is calculated and shown in the last column of Table 6.4. The Cross Emergence Function and Index of Divergence for five countries of Section (B) are calculated for the entire duration of the 166 weeks.

<u>TABLE 6.4.</u>   **THE 166 WEEKS OF COVID-19**
         CROSS EMERGENCE FUNCTION AND INDEX OF DIVERGENCE

WHO REGION	Countries	Cases	ζ_0	$H(g\|x)$	$H(x\|g)$	$\mathcal{E}(g\|x)$	$\mathcal{E}(x\|g)$	$\Delta\mathcal{E}$
W PACIFIC	China	99 m	129.6	189.15	75.71	17313	9815	**7498**
W PACIFIC	Korea	30.5 m	123.3	81	33.24	7414	4099	**3315**
EUROPE	Germany	38.13 m	100	41.17	30.32	3768	3032	**736**
EUROPE	Spain	13.83 m	79.4	44.48	35.62	4071	2827	**1244**

| AMERICAS | USA | 102.25 m | 78.4 | 40.37 | 38.27 | 3695 | 3002 | **693** |
| GLOBAL | | 758.5 m | 91.5 | - | - | - | - | - |

6.10. Summary. The fractal nature of COVID-19 pandemic has been identified in the analyses presented in the present chapter and the previous ones. The scale invariance and the self-similarity of the pandemic at local and global scales, for the two-time scales, were demonstrated. The emerging pandemic have been treated as the dynamical system that is sensitive to the initial conditions thus the growth of the population of the Coronavirus infected persons in slightly different initial conditions can produce quite different outputs. The sensitivity to the initial environmental and spatial constraints is also displayed in the varying nature of the waves of the infected persons' cases and stages of the pandemic identified through the diverging trends of emergence of the cumulative cases as a function of the days of the pandemic.

Three significant aspects of the COVID-19's dynamical emergent and cross-emergent system have been described through the Emergence Functions, cross entropies, Cross Emergence Functions, and the Indices of Divergence.

1. The Emergence Functions λ demonstrated the emerging trends of the chosen countries as well as the cumulative global emergence. The relevance of λ depends on its descriptive quantification of the temporal emerging trends of the cases through the inflexion point ζ_0 and the entropic parametrization of the pandemic through H_m, demonstrated in Figure 6.8. The individual as well as the global λs were reported earlier, here, their ratios are compared. The Emergence functions λ for the initial 100 days of the pandemic, derived from the data presented in Figures 6.1 to 6.3, were

tabulated in Table 6.1 and for the 166 weeks' duration in Table 6.3.

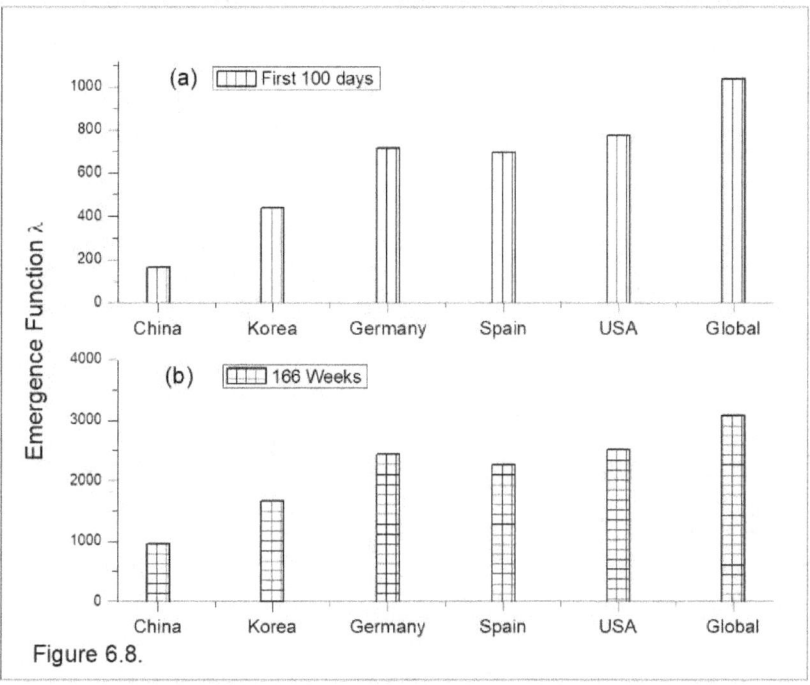

Figure 6.8.

In Figure 6.8, the Emergence Function λ is plotted as a function of the country. The upper graph for the initial 100 days. It shows that after the initiation of the pandemic, the Chinese Emergence function showed a marked divergence from the global emergence, whereas the initiation of COVID-19 in Germany, Spain and USA closely seem to follow the global trends of the emergence of Coronavirus. The divergence in the case of Korea of the ratio λ(g)/λ(x) is less than that of the Chinese and higher than the three countries representing the regions of Europe and the Americas. The lower graph for the 166 weeks describes the similar trend, as of that during the initial 100 days, with lesser divergence for the Chinese and the Korean cases: the remaining three countries present similar profiles as that during the initial spread.

2. The ratios of Cross Emergence Functions $\mathcal{E}(g|x)$ and $\mathcal{E}(x|g)$ in Figure 6.9 are derived from the data was tabulated above in Table 6.2 for the initial 100 days and in Table 6.4 for the cumulative 166 weeks of the pandemic. Except for the initial 100 days of the pandemic in China, the ratios

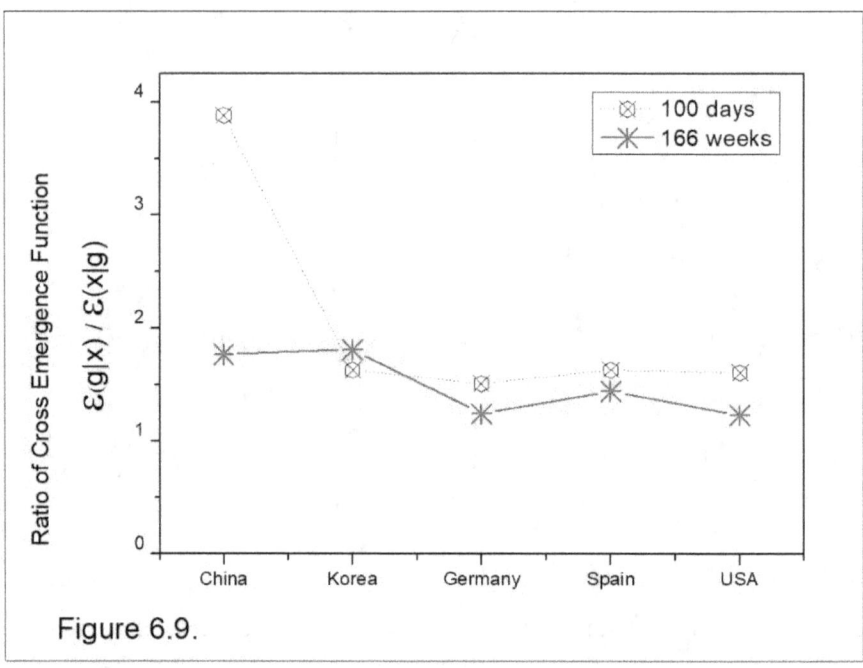

Figure 6.9.

of the CEFs present an approximately coherent picture of the cross emergent COVID-19. Cross Emergence Function provide a relative measure of the pandemic as it emerges and spreads in different countries, regions and globally. The shorter duration of the initiation of the pandemic (100 days) is shown to lead to relatively higher divergences in the patterns and profiles of the spread of the virus among the different populations. The longer durations seem to have created similarities in the emergence patterns in diverse countries and environments. Keeping in view the observed dissimilarities in the handling of and dealing with COVID-19, the duration of the pandemic extended to more than three years seem to have led to similar cross emergent profiles.

3. The Indices of Divergence ΔƐs for the five countries, tabulated in Tables 6.2 and 6.4, evaluated as the absolute value of the difference between CEFs Ɛ(g|x) and Ɛ(x|g), are plotted for each country in Figure 6.10. The two graphs for the 100 days and 166 weeks' duration, validate the observations made above. It was pointed out in Figures 6.8 to 6.10 that there were significant divergences in terms of Emergence and Cross Emergence Functional evaluation of the Western Pacific and Europe and Americas. It can also be noted that the shorter and longer periods of pandemics in the individual countries and globally, present approximately similar order of magnitude Indices of Divergence ΔƐ.

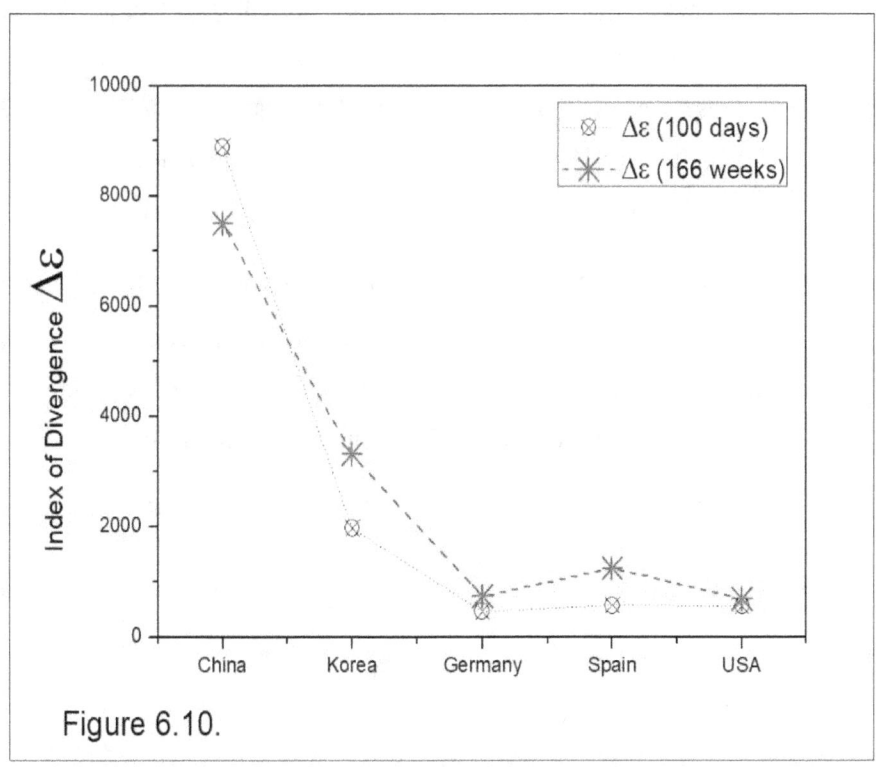

Figure 6.10.

Graphical Summary. The Cross Emergence Functions and the IoDs, derived from the Probability distributions of the COVID-19 cases in the five chosen countries of the three(3) WHO-designated Regions, against the Global distributions are evaluated in this Chapter. From the data for CEFs in Tables 6.2 and 6.4, the bar graphs for the initial 100 days and the cumulative 166 weeks of the pandemic are shown in (a) and (b), for the USA, Germany, Spain, Korea, and China.

The Indices of Divergence indicate close resemblance between the two durations of the Pandemic. Apart from self-similarity and scale invariance, the Pandemic's emergence profile demonstrates remarkable temporal independence.

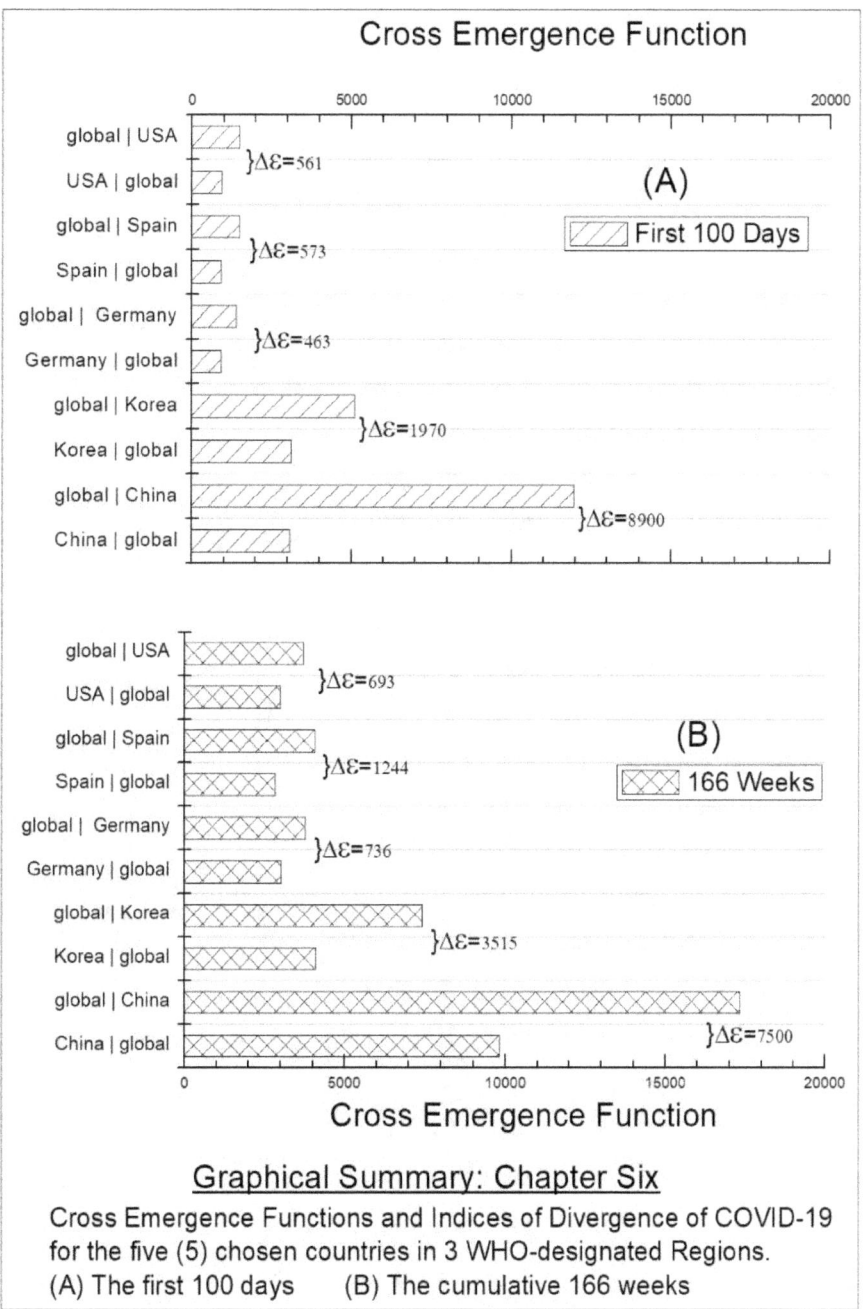

Graphical Summary: Chapter Six

Cross Emergence Functions and Indices of Divergence of COVID-19 for the five (5) chosen countries in 3 WHO-designated Regions.
(A) The first 100 days (B) The cumulative 166 weeks

CHAPTER 7

EVOLUTION OF SARS-COV-2 IN THE USA

7.1. USA as the model for studying the evolution of SARS-CoV-2. The emergence and cross emergence of the SARS-CoV-2 in the USA has been investigated in detail using the COVID-19 data on the infectious cases in Chapter Six. The data was inclusive of all the infected cases, but without distinguishing between the type of the variant causing the infections. Here, in this Chapter, the evolution of the severe acute respiratory syndrome coronavirus-2 is investigated by identifying the source of the infections causing illness. The data available from the sequencing of the variants, in the USA, during the specific period of just over two years, will be utilized to infer the relative share of the evolving variants, in the increasing population of the COVID-19 cases, by employing the diagnostic tools of Emergence Function λ, Cross Emergence Functions $\mathcal{E}(p|q)$ and $\mathcal{E}(q|p)$, where p and q represent the parent-offspring or the sisters variants, to be discussed in detail later in this chapter. The IoDs $\Delta\mathcal{E}$ will be employed, as in the earlier chapters, to provide the measure of divergence between the cross-emergent variants.

The data available on the variants of SARS-CoV-2 for the USA will be analyzed using the information-theoretic model developed and applied in the previous chapters. The USA is chosen for the scrutiny of the spread of the pandemic with the help of the biweekly data of the results of the sequences of the SARS-CoV-2 in the USA. The data is available on the web site of **ourworldindata.org** [85,86] that has been used in this chapter to construct the probability distribution functions from the relative number densities of each new variant. The starting date of the biweekly sequences is 1 March 2021 which is ~60 weeks after the global onset of COVID-19 in early 2020.

USA had the largest number of cases that were monitored with high precision and the most appropriate response in terms of the provision of health care to the infected persons, monitoring of the emerging scenarios, and by providing transparent information of the pandemic to the WHO. The WHO data of the infected cases, during the temporal space of 166 weeks, has been utilized in the pandemic's information-theoretic description in this book. Chapter Five was dedicated to studying the emerging pandemic globally, and in the six (6) WHO-designated regions by treating COVID-19 as an emerging, cross emerging, and self-organizing dynamical system. In Chapter Six, the distinctive and diverging emergence profiles of the coronavirus pandemic in the five countries representing the three major regions-the America, Europe, and Western Pacific, were analyzed in detail. The USA was studied as one of the five countries.

To understand the evolution of the SARS-CoV-2 through the mutations of the virus into other forms that are higher in transmissibility and immune escape, the data on the biweekly sequencing of the infected patients provides the essential clues. From the perspective of the virus, its survival and further transmission depend on the ability to evolve transmission-enhancing characteristics [87-93]. The enhancement in the intrinsic transmissibility with the elevated infectiousness leads to the mutations that are more effective in immune escape [refs. 93-103 and the references therein].

In Chapters 4 to 6, the Index of Divergence, defined as the absolute value of the difference of the two Cross Emergence Functions denoted as $\Delta\mathcal{E} = |\mathcal{E}(p|q) - \mathcal{E}(q|p)|$, was used to identify the magnitudes of the dissimilarities of emergence of the pandemic in different countries and regions. Here, the Index of Divergence will be used to identify the divergent/emergent characteristics of the emerging Variants of Concern-the VOCs by analyzing the data for the mutating viruses.

The information-theoretic model was applied to the emergence and evolution of SARS-CoV-2 in two distinct categories; (1) to the emergence of the infectious cases locally and globally in Chapters 4 to 6 by treating the global pandemic as an information-generating dynamical system and (2) to analyze the dynamics of the evolving variants in the USA in the cross-emergent analyses, here, in this Chapter. These aspects provide the distinction between the predictive aspects of some of the mathematical models [104-116] and the one presented in this book. This distinction will be further highlighted in the next Chapter Eight.

7.2. Evolution of the variants of SARS-CoV-2 in the USA. In Figure 7.1(a), the number of the weekly cases of COVID-19 are shown as these emerged throughout the USA. For the sake of comparison with the analyses presented in Chapter 6, it is pointed out that the present figure is to describe the significant changes in the accumulating virus infections induced by the evolving, mutating SARS-CoV-2, the source of COVID-19. The last chapter presented the analyzed data for the cumulative cases, without distinguishing and identifying between the cases due to the different variants in the USA.

Figure 6.6 of Chapter 6 had the profile of the pandemic in USA over the extended duration of 166 weeks. The last chapter's analyses of the emergence of the cases were done as the cumulative sum of the successive stages of the COVID-19 without distinguishing between the mutating SARS-CoV-2 variants. Following the emergence of the pandemic in early 2020, after ~8 months, the divergent SARS-CoV-2 lineages were started being detected, sequenced, and reported in various parts of the world [87-103]. These early lineages were coded as Variants of Interest (VOIs) and the later ones were termed as the Variants of Concern (VOCs).

The first, and most significant, set of VOCs were termed Alfa, Beta, Gamma, and Delta. Other variants were also identified. Our focus is on the emergence of the divergent lineages (Alfa, Beta, Gamma, Delta) that later led to the within-lineage evolution of the Omicron series of the interlinked lineages [89-91]. In November 2021, the earliest batch of the lineages termed BA.1, BA.2 led to Omicron sub-lineages. BA.1 was the dominant variant that emerged and followed the Delta VOC, globally as well as in the USA.

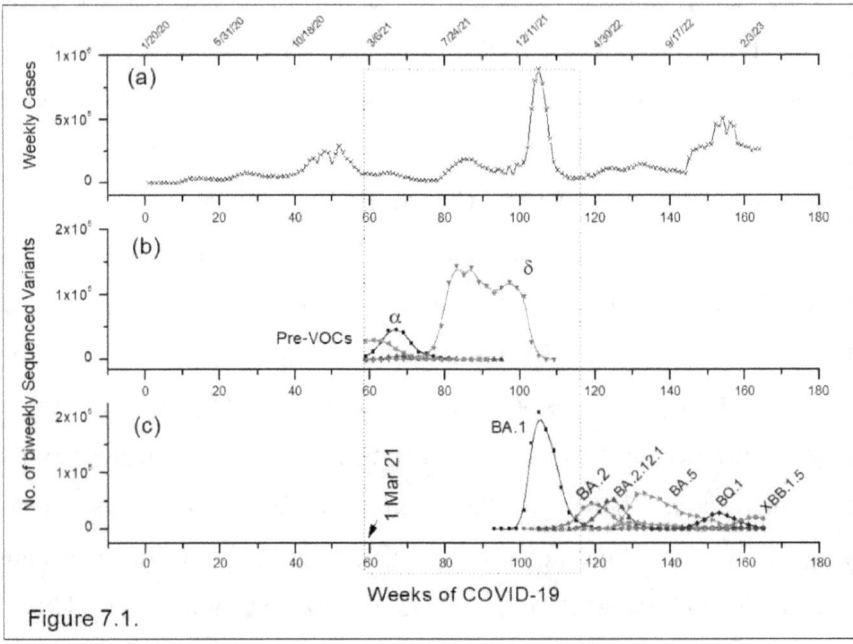

Figure 7.1.

Figure 7.1. Variants of COVID-19 in the USA. (a) More than 100 million COVID-cases are shown in the form of the weekly cases over the course of ~166 weeks. The format of the weekly cases is to elaborate the contribution of the mutating variants of the virus in the next two figures. (b) The ratios and number of the biweekly analyzed sequences of the SARS-CoV-2 Variants of Concern (VOCs) are shown. The ratios were obtained from approximately ~ 1.5 million biweekly analyzed sequences. (c) The Omicron variants are plotted as these emerged and transmitted throughout the USA. The first, massive peak is due to the Omicron variant BA.1 that started around the 100th week of the pandemic. It was the major contributor to USA cases in early 2022. The 2.5 million biweekly sequences identified the series of evolving variants from BA.1 onwards.
Data from https://ourworldindata.org/grapher/covid-variants-bar [85,86].

Figure 7.1(b) plots the share of the biweekly analyzed sequences of the Variants of Concern-VOCs that followed the Pre-VOCs. The number densities of the Pre-VOCs show the relative share of the SARS-CoV-2 variants that preceded the Early VOCs. The Early VOCs consisted of the Alfa, Beta, Gamma, and Delta starting from 1st of March 2021 for the next ~

40 weeks period to the end of 2021. Figure 7.1(b) shows the transition where the number densities of the Pre-VOCs continuously decreased while the emerging trends of the Early VOCs can be identified.

The Early VOCs emerged, existed for a certain period, mutated, and then gave way to the next phase of the emerging variants-the Omicrons. The increasing number densities of the first of the Omicron variant-BA.1 can be seen to follow at the stage when the numbers of the Delta variant start to decrease. These introduced a new phase where the multiple Omicron sub-lineages became the globally dominant VOCs, as shown in Figure 7.1(c)

7.3. The Emerging, Evolving SARS-CoV-2. Figure 7.2 is reconstructed from the two segments of Figure 7.1(b) and (c). It contains the normalized ratios of the emerging variants of SARS-CoV-2 from the data of the biweekly sequences [86]. The two-year data identifies the stages when each variant evolves, spreads with its specifically acquired transmissibility rate that demonstrate the enhanced acquired characteristics of the variants. Multiple variants could be operating at a given time and place. This can be observed in Figure 7.2. The figure divides the evolution profile in two broad phases. The first phase consists of the stage where the four of the Early VOCs named Alfa, Beta, Gamma, and Delta emerged, replacing the diminishing cases due to the earlier form of the virus identified as Pre-VOCs.

The first phase of the 45-50 weeks from 1 Mach 2021 onwards, displays the relative ratio of the declining numbers of the Pre-VOCs infected cases and the gradually increasing ratios of the cases due to the Early VOCs classified as Alfa, Beta, Gamma, and Delta. The cases identified in the biweekly sequences with the Delta variant can be seen to rise sharply within the first ~10 weeks.

The Early VOCs dominated temporal space in the biweekly sequences also contain a total of 0.15M cases infected with the Pre-VOCs. It followed by the 0.23M cases due to the Alfa, 0.03M to Beta, 0.028M to Gamma, and ~1.5M cases due to Delta VOC. The Delta variant dominated all other variants (α, β, γ) after the first 20 weeks. It dominated for almost half a year from June to December 21, until the Omicron variants emerged with higher transmissibility and replaced the Delta VOC.

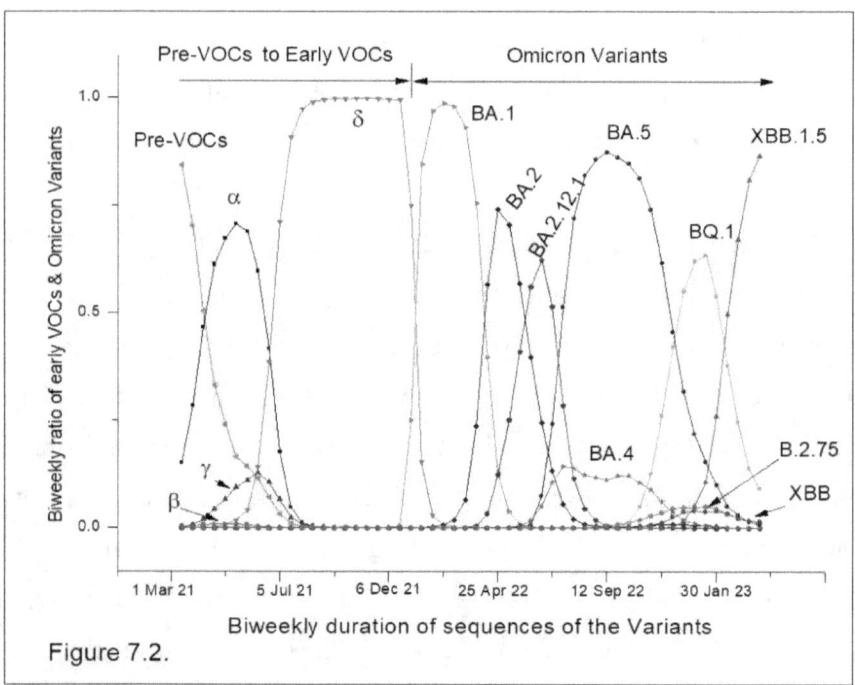

Figure 7.2.

Figure 7.2. The two phases of the ratios of Pre-VOCs, the Early VOCs, and Omicrons in the USA. The results of the normalized ratios of the Variants of Concern (VOCs) are shown from the biweekly sequences of the virus, starting from 1 March 2021. The emergence of the evolving Variants of Concern (VOCs) was analyzed in ~4 million sequences in the USA over ~115 weeks.

The second phase of Figure 7.2 shows the relative ratios of series of Omicron VOCs starting with BA.1 in early November 21. Within the next 20 weeks it replaced the Delta-variant. The series of the sister lineages BA.1, BA.2….BA.5 evolved and remained the major source of infection during the year 2022. Onwards from late 2022, the new, sequences of the emergent variants BQ.1, XBB, and XBB.1.5 can be seen.

The evolution of SARS-CoV-2 continued in the years 2023-4. The WHO and various national and international COVID-19 monitoring agencies continued to monitor, diagnose, and report the data on the evolving variants and the infected cases due to their sub-lineages [80,86].

7.4. Information-theoretic profiles of the Early VOCs. The information-theoretic analysis of the first stage of the sequences of the SARS-CoV-2 variants is shown in Figure 7.3(a) as the instantaneous entropic $[p(\zeta) * \ln(1/p(\zeta))]$ plots of the Pre-VOCs and the Early VOCs Alfa, Beta, Gamma, and Delta. The graphs are plotted as a function of the biweekly sequences conducted and reported from 1 March 2021 onwards. In Figure 7.3(a), the diminishing sequences of Pre-VOCs and the emerging stages of the Alfa, Beta, Gamma, and Delta VOCs occur during approximately the same time-zone. Following the first graph for the cases that showed Pre-VOCs infections, the series of the cases due to the emerging variants of SARS-CoV-2 were recorded. The SARS CoV-2 mutations yielded the more transmissible, vaccine-evading forms of the VOCs [87,88].

The noticeable feature of Figure 7.3(a) is that VOCs Alfa and Beta both had significant presence on 1 March 2021, implying that these variants coexisted with the Pre-VOCs earlier than 1 March when the first reported result of the biweekly sequencing was reported. The emergence of the Gamma and Delta variants followed when number densities of Alfa and Beta had started to decrease. Gamma VOC evolved, emerged, and then diminished within the next ~25 weeks. The Delta variant had a longer ~45 to 50 weeks' existence in the temporal zone of the SARS-CoV-2 and dominated all others.

The cumulative entropic plots of Figure 7.3(b) describe the information profiles generated by the four reported VOCs and the Pre-VOCs. These graphs represent the characteristics of the variants from their emergence to decline. Their coexistence, in the varying ratios, allows comparative analyses by using the cross entropies of the pairs of the chosen variants.

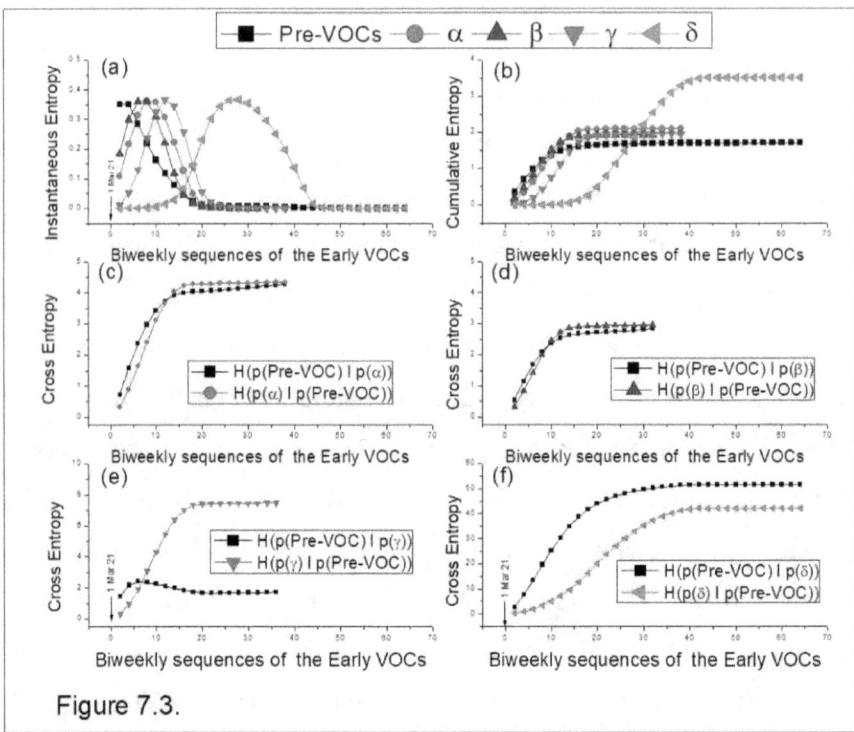

Figure 7.3.

Figure 7.3. The instantaneous, cumulative, and cross entropies of Pre-VOCs and the Early VOCs. (a) The instantaneous entropic plots of $[p(\zeta) * \ln(1/p(\zeta))]$ plotted as a function of ζ-the biweekly duration of sequencing of the variant. Here, the probability distributions $p(\zeta)$ represent the probabilities of $\alpha, \beta, \gamma, \delta$ and the Pre-VOCs. (b) The cumulative entropic plots present the respective profiles of emergence of the cases. The initial growth curves settle into constant values when the variant stops evolving. (c) The two Cross entropic graphs for the Pre-VOCs and Alfa are drawn between the probability distributions $p(\alpha)$ and $p(\text{Pre} - \text{VOCs})$. (d) The Cross entropic graphs for the probability distributions $p(\beta)$ and $p(\text{Pre} - \text{VOCs})$. (e) The Cross entropic graphs for the probability distributions $p(\gamma)$ and $p(\text{Pre} - \text{VOCs})$. (f) The Cross entropic graphs for the probability distributions $p(\delta)$ and $p(\text{Pre} - \text{VOCs})$.

Figure 7.3(c) to (f) plot the cross entropies of the four Early VOCs with the Pre-VOCs. Cross entropy sets of the twin probability distribution functions

p(True) and p(Expected) are being used as a measure of the divergence between the distributions. Cross entropy is calculated by assigning p(True) to p(Pre − VOCs) and p(VOCs), alternatively. Four sets of the eight cross entropies are displayed in Figure 7.3(c) to (f) by alternating the distributions $p(\alpha)$, $p(\beta)$, $p(\gamma)$, and $p(\delta)$ as p(True) and p(Expected) with p(Pre − VOCs).

The two cross entropies of the probability distributions p(Pre − VOCs) and $p(\alpha)$ show low divergence. $H[p(\text{Pre} - \text{VOCs})|p(\alpha)] \sim H[(p(\alpha)|p(\text{Pre} - \text{VOCs})]$ in Figure 7.3(c). Similarly, the Pre-VOCs and Beta generate approximately equal cross entropies towards each other. These indicate negligible divergence between their cross entropies.

The Gamma variant in Figure 7.3(e) displays initially a smaller cross entropic divergence that enlarges to a wider separation among the distributions of Gamma and Pre-VOCs. A factor of 2 larger cross entropic divergence occurs between $p(\gamma)$ and $p(\text{Pre} - \text{VOCs})$ as compared with those in Figure 7.3(c) and (d). Figure 7.3(f) shows that the cross entropies for the δ variant $H(p(Pre - VOCS)|p(\delta)$ and $H(p(\delta)|p((Pre - VOCS))$ during its emergence, start to display a continuously increasing cross entropic divergence. The divergence increases to an order of magnitude higher than the corresponding ones for Alfa, Beta, and Gamma VOCs.

Table 7.1 tabulates the relevant inflexion points ζ_0, and the Emergence Function λ for the Early VOCs that emerged during the later diminishing phase of the Pre-VOCs. The cross entropies and the Cross Emergence Functions calculated for each of the pairs of the probability distributions of Pre-VOCs denoted as p and the Early VOCs shown with their corresponding distributions q. The Index of Divergence $\Delta\mathcal{E} = |\mathcal{E}(p|q) - \mathcal{E}(q|p)|$ are calculated and shown in the last column. In the third column of Table 7.1, the sequence of the inflexion points ζ_0 start with the 6.23 weeks for Pre-

VOCs, then follows Beta, Alfa, Gamma, and Delta. Delta has the longest emergence profile with $\zeta_0=31.6$ weeks. Its emergence coincides with the Pre-VOCs and the three Early VOCs after emergence and its diminishing phase with the emerging stage Omicron BA.1 which is the first of the Omicron VOCs.

TABLE 7.1. THE EMERGENCE AND CROSS EMERGENCE FUNCTION AND INDEX OF DIVERGENCE OF THE EVOLVING VARIANTS WITH REFERENCE TO THE PRE-VOCS

PDF	Variant	ζ_0	λ	$H(p\|q)$	$H(q\|p)$	$\mathcal{E}(p\|q)$	$\mathcal{E}(q\|p)$	$\Delta\mathcal{E}$
$p \equiv$	Pre-VOCs	6.23	10.2					
$q \equiv$	α	9.25	19.2	4.25	4.35	43.16	83.29	**40.13**
	β	8.4	15.96	2.8	2.95	28.43	47.08	**18.65**
	γ	11.8	22.8	2.4	7.5	24.37	170.8	**146.4**
	δ	31.6	111.6	51.5	42.2	522.9	4707	**4184**

The Emergence Function λ in the 4th column has the lowest value for the Pre-VOCs. This is understandable as the data for the complete evolutionary stages of the Pre-VOCs is not shown here. The low value of λ has been calculated from the probability distribution for the number of cases during its diminishing stage. The values for the Emergence Function λ of the Early VOCs increase from ~16 for Beta, to 19.17 for Alfa, 22.8 for Gamma and 111.6 for the Delta VOC.

Table 7.1 displays the CEFs $\mathcal{E}(p|q)$ and $\mathcal{E}(q|p)$ constructed from the inflexion points and the cross entropies $H(p|q)$ and $H(q|p)$ by alternatingly treating the probability distributions $p(\text{Pre}-\text{VOCs}) \equiv p$ and $p(\text{Early VOCs}) \equiv q$. The sets of CEFs $\{\mathcal{E}(p|q), \mathcal{E}(p|q)\}$ for the four combinations of Pre-VOCs and the Early VOCs generate the Indices of Divergence $\Delta\mathcal{E}$ tabulated in the last column. The numerical value of $\Delta\mathcal{E}$ is minimum for {Beta-Pre-VOCs} and maximum for the set {Delta-Pre-

VOCs}. The value of $\Delta\mathcal{E}$~4200 for the distributions of Delta and Pre-VOCs is the highest amongst the Indices of Emergence for all the VOC. This identifies the variant that evolved with the highest relative transmissibility amongst the Early VOCs

7.5. Cross Entropic divergence within the VOCs. The profile of emergence from the Pre-VOCs towards the Early Variants of Concern, are again displayed in Figure 7.4(a). The objective is to put this important stage of the evolution of SARS-CoV-2 in perspective. The four variants Alfa, Beta, Gamma, and Delta emerged from and replaced Pre-VOCs that continued to circulate without major mutations for almost ~60 weeks after the emergence of COVID-19 in the USA. Multiple mutations that increase transmissibility and immune escape lead to the sequence of emergence of the VOCs with larger CEFs. Alfa and Delta variants became more significant with their evolved characteristics and had greater impact in terms of the increased number of infections over the next year before the emergence of the Omicron series of VOCs.

Figure 7.4(b) sums up the resulting cross entropies of the four emergent VOCs with Pre-VOCs plotted in the same graph. These are already present as separate figures in Figure 7.3(c) to (f). Here, Figure 7.4(b) aims to demonstrate the relative magnitudes of the cross entropic divergences to clearly show that; (1) Alfa and Beta variants had minimal cross entropic divergence with the Pre-VOCs, (2) Gamma demonstrated a factor of four higher divergence against Pre-VOCs than the corresponding one for the Pre-VOCs to Gamma implying $H[p(\text{Pre} - \text{VOCs})] \sim 4 * H[p(\text{Pre} - \text{VOCs}) \mid p(\gamma)]$, and (3) Delta and Pre-VOCs display the largest divergence ~ 40 and 50 cross entropic units.

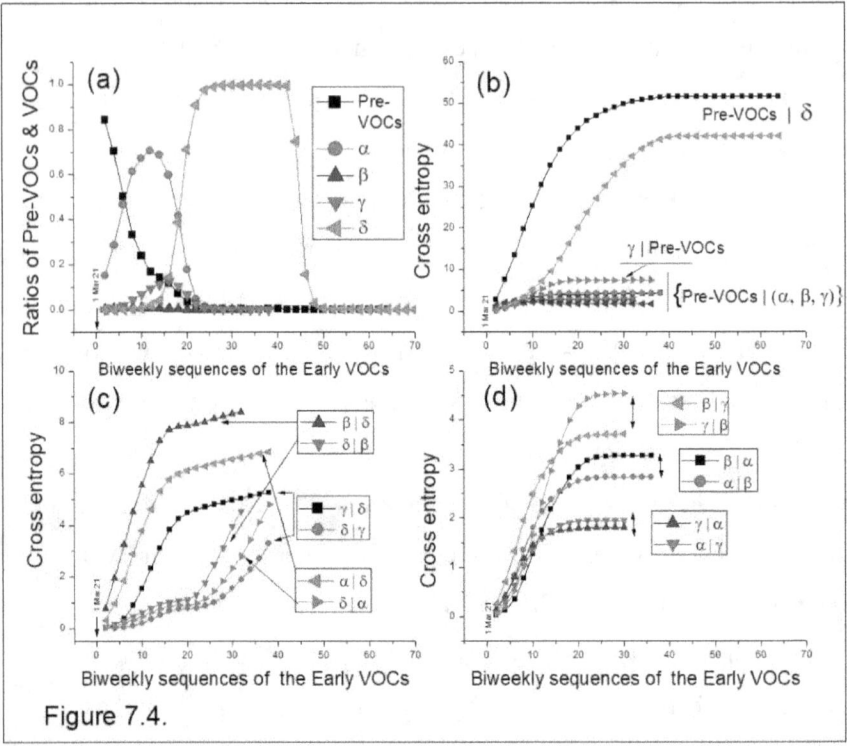

Figure 7.4.

Figure 7.4. Cross entropies of the Early VOCs. (a) The normalized ratios of the cases of emerging profiles of the Early VOCs are shown with the emphasis on their relative emergence, reaching a peak, and decline. The Pre-VOCs are shown here as the benchmark of extinction. (b) Starting from 1 Mar 21, the four sets of the twin cross entropic profiles are plotted as a function of the weeks of COVID-19. The two early variants Alfa and Beta demonstrate similar patterns of cross entropies with the Pre-VOCs. The Delta variant demonstrated considerably higher divergence from the Pre-VOCs. (c) and (d) The cross entropies of the combinations within the VOCs. All graphs are plotted against the biweekly sequencing stages of the VOCs.

The cross entropies amongst the VOCs are calculated and shown in Figure 7.4(c) and (d). Six sets of the twin cross entropies are plotted for the combinations of the VOCs Alfa, Beta, Gamma, and Delta. The twin cross entropic combinations with Delta are shown in Figure 7.4(c) with the designated symbol of $\gamma|\delta \equiv H((p(\gamma)|\,p(\delta)))$. Similarly, $\delta|\gamma \equiv H((p(\delta)|\,p(\gamma)))$. The same notation is used for the other two sets of graphs with cross entropic symbols shown in the figure for α, β. The maximum cross entropic divergences are for the distributions for the variants β and δ. The probability distributions for α, β and γ demonstrate a factor of two lower cross entropic distances among themselves in Figure 7.4(d).

TABLE 7.2. CROSS EMERGENCE FUNCTION AND INDEX OF DIVERGENCE AMONGST THE EARLY VOCS

VARIANT (p)	Variant (q)	$\mathcal{E}(p\|q)$	$\mathcal{E}(q\|p)$	$\Delta\mathcal{E}$
α	β	37.2	28.8	8.4
β	γ	59.2	103.1	44
γ	α	54.3	74.6	20
δ	α	131.5	536.5	405
δ	β	134.6	506.4	372
δ	γ	120	435	315

Table 7.2 has the Cross Emergence Function and the Index of Divergence for the six sets of the four VOCs α, β, γ and δ calculated from the data in Figure 7.4.

7.6. The transition from Delta to the BA VOCs. This section starts with the decaying numbers of the most significantly evolved member of the Early VOCs-the Delta variant, and the emergence of the evolving BA variants. Figure 7.5 plots the emergence profile of the Omicron series of SARS-CoV-2. The instantaneous entropies of the nine selected VOC are plotted as a function of the biweekly sequencing of each of the emerging variant. Figure 7.5(b) has the cumulative entropic graphs of the Omicrons from BA.1 to XBB.1.5. These graphs describe the information theoretic footprints of the variants' emergence, their spread with relatively increased levels of transmissibility, to be gradually replaced by the next higher level transmissible variant's emergence.

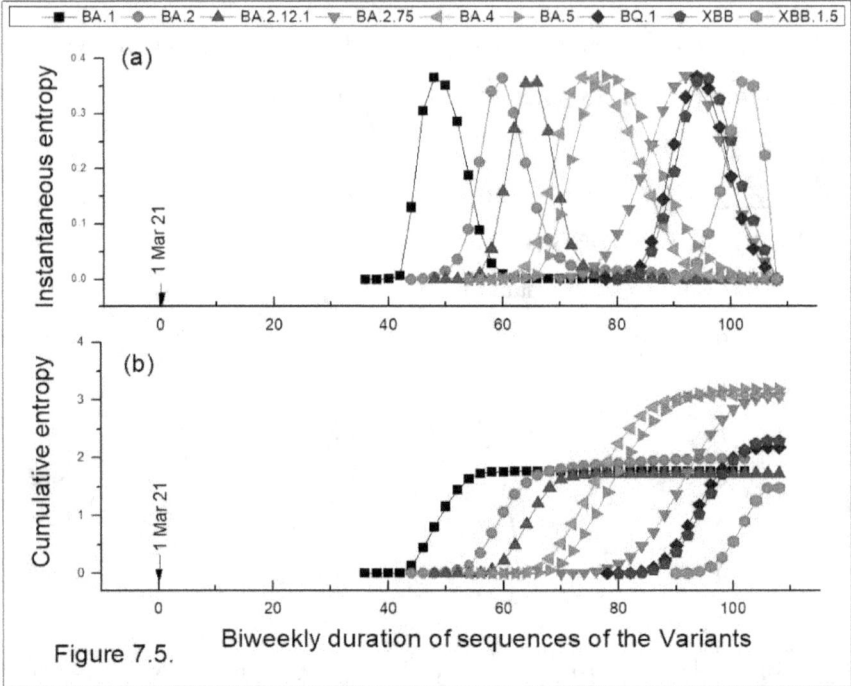

Figure 7.5. The emergence profile of Omicrons. (a) The instantaneous entropic plots as a function of the biweekly sequencing of the emerging variants. (b) The cumulative entropic graphs of the Omicrons from BA.1 to XBB.1.5 describe their entropic footprints, as these variants emerge and diminish.

The Emergence Function λ for the nine (9) Omicron variants are constructed and tabulated in Table 7.3 by utilizing the entropic profiles of the Omicron VOCs. The nine SARS-CoV-2 variants were dominated by the BA lineages from BA.1 to BA.5 with varying degrees of transmissibility. The table below shows that BA.5 had the highest Emergence function $\lambda \sim 80$ followed by BA.4 and BA.2.75. It must be pointed out that the later versions i.e., XBB and XBB.1.5 can be seen to be still spreading during the last stages of the biweekly sequences reported up to the 10th of April 2023. Therefore, the initiation of the Omicrons series with BA.1 can be evaluated more comprehensively as is done in Chapter. The final concluding remarks and analyses of the ongoing processes of the mutating virus can be best attempted when the pandemic stops spreading.

TABLE 7.3. EMERGENCE FUNCTIONS λ OF THE OMICRONS

VOCS	ζ_0	λ
BA.1	12.12	21.5
BA.2	15.7	31
BA.2.12.1	15.64	27.7
BA.2.75	20.8	64.5
BA.4	21.74	~67
BA.5	25.14	80.5
BQ.1	15.52	33.8
XBB	14.72	33.7
XBB.1.5	9.06	13.4

Figure 7.6 deals with the two sets of VOCs, in pairs, that shared a similar temporal space. The first set comprises Delta and the emerging Omicron BA.1 in Figure 7.6(a), (c) and (e). The pair of the instantaneous entropies are plotted in Figure 7.6(a) while (c) and (e) have the consequent cumulative entropies and the cross entropies. Whereas, the cumulative entropies are due to the individual variants, the cross entropies are evaluated with respect to each other, treating alternatingly the one as the true and the other as expected. Figure 7.6(b), (d) and (f) have the same entropic profiles for the Omicron variants' pair of BA.1 and BA.2.

The distinctive features from the information-theoretic characteristics of the two pairs of variants the (Delta, BA.1) and (BA.1 and BA.2) are as follows.

1. The cumulative entropies, plotted in 7.6(c) of the Delta and BA.1 pair are distinctively separated from each other.

2. The cumulative entropies of BA.1 and BA.2 are initiated from and within the same time zone and display the information profile of a diminishing variant with an emergent one. Initially, BA.1 generates higher entropy from the start until the 50th week and stays constant thereafter, implying no significant information generation. The cumulative entropy of BA.2 grows slow initial growth after the evolution of BA.2 variant, reaches to the entropy equalizing stage at ~60 weeks with that of BA.1 and continue increasing thereafter as the new variant.

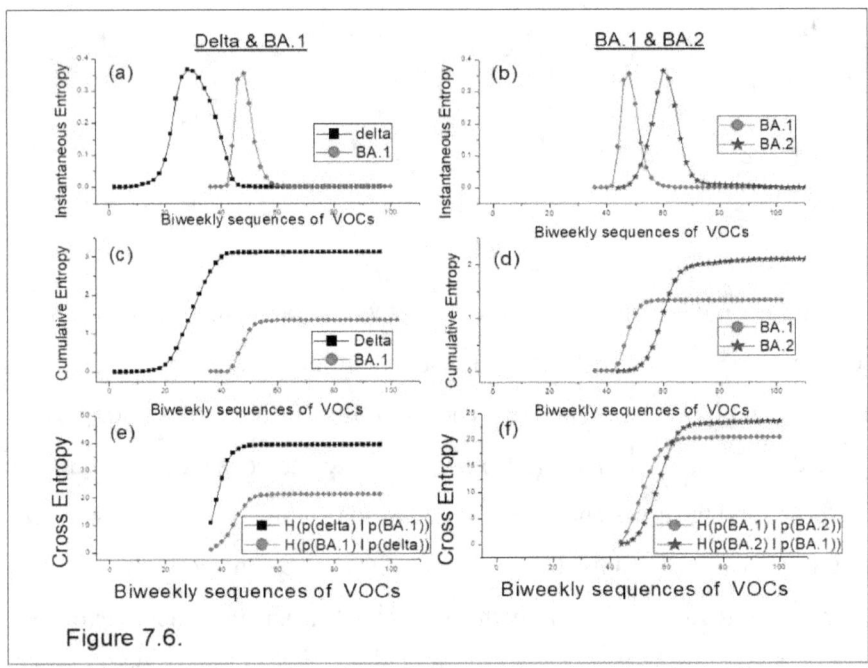

Figure 7.6.

Figure 7.6. The transition from the Delta VOC to the BA Omicrons. (a) and (b) The instantaneous entropic plots as a function of ζ-the biweekly duration of sequencing of the variants. The probability distributions $p(\zeta)$ represent the probabilities of δ and the BA.1 in (a) and probabilities of BA.1 and BA.2 in (b). (c) The cumulative entropic plots of δ and the BA.1 describe the respective profiles of their divergent emergence. (d) The cumulative entropies of BA.1 and BA.2 show their emergence in a similar time zone. (e) The two Cross entropic graphs for the Delta and BA.1 demonstrate the divergent VOCs (f) The Cross entropic graphs for the probability distributions $p(BA.1)$ and $p(BA.2)$ show the two sister lineages displaying the same dynamic traits.

3. The cross entropies of the Delta and BA.1 pair demonstrates that starting with the difference of ~8 units at the beginning, the difference of the two cross entropies reaches ~20 cross entropic units in the next 20 weeks.

4. In Figure 7.6(f), the BA.1 and BA.2 pair of the Omicron VOCs starts from almost equal and minimum cross entropies and reaches to ~20 cross entropic units at the stage when the two cross entropies equalize.

7.7. The emergence profiles of the within-lineage evolution. Let us look at the emergence profiles of the three of the Omicron VOCs BA.2, Ba.2.12.1 and BA.5. These were among the dominant VOCs during the period starting from 6 December 21 when the Omicron family of VOCs emerged replacing the Delta variant.

The two Omicron variants of the sister lineage BA.2 and BA.2.12.1 in Figure 7.7(a) emerged almost within the same temporal space and grew within few weeks to their respective maxima separated by ~8 weeks. Their cumulative and the cross entropies in Figure 7.7(c) and (e) present similar emergence profiles. The maximum cross entropic divergence is ~1.5. The variant BA.2.12.1 shows moderate divergence from BA.2 in emerging as a distinct variant.

Figure 7.7(b) shows the declining BA.2.12.1 variant and the emerging BA.5. In Figure 7.7(d), the crossover of BA.5 occurs around the 80th week.

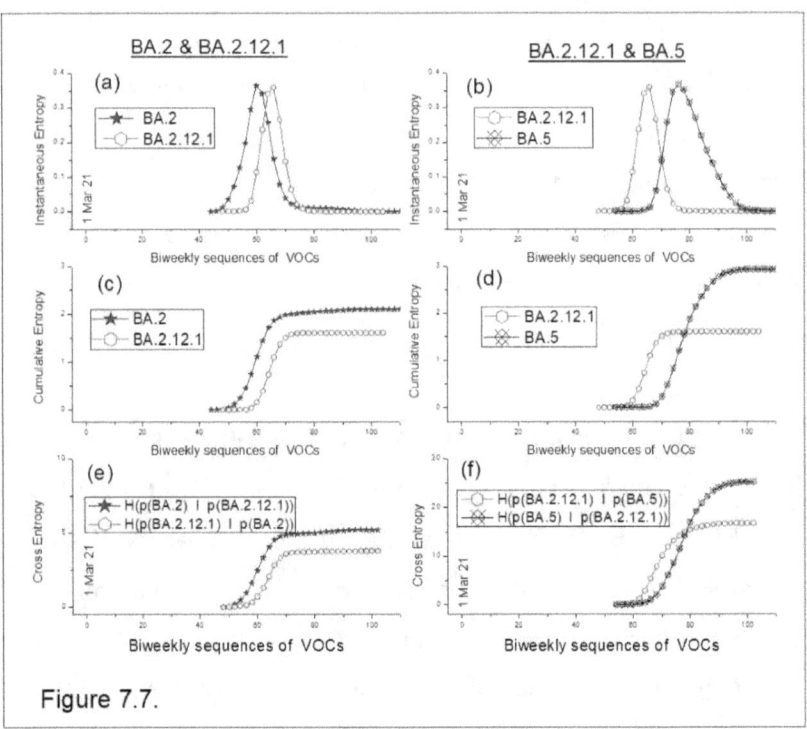

Figure 7.7.

Figure 7.7. The profiles of the within-lineage evolution of BA lineages. (a) The instantaneous entropies of the BA.2 and the evolved sub lineage BA.2.12.1. (b) The instantaneous entropic graphs BA.2.12.1 and BA.5. (c) The two entropic graphs of BA.2 and BA.2.12.1 show divergent profiles. One for the diminishing variant BA.2 and the other for the evolving BA.2.12.1. (d) The cumulative graphs of BA.2.12.1 and BA.5 show the cumulative entropies. (e) Cross entropies of BA.2 and BA.2.12.1 depict the earlier entropic profile, shown in (c), of the divergence of the two variants. (f) The cross entropic graphs of BA.2.12.1 and BA.5 show the two variants sharing similar characteristics demonstrated through their probability distributions and cross entropies.

It took BA.5 almost ~25 weeks from emergence to equalize entropy with that of BA2.12.1 and to continue to grow, while BA.2.12.1 had generated the maximum information about its evolution and then being replaced. The difference of ~10 units in the set of the twin cross entropies generated by the probability distributions of BA.2.12.1 and BA.5 demonstrate the diverging trends that also indicate the emergence of BA.5 as a major evolving variant with higher transmissibility as compared with its predecessors.

Figure 7.8 has six graphs of the cross entropies only. As has been indicated earlier in this chapter and elsewhere that the net difference between the values amongst the sets of the twin cross entropies of any of the two variants, that share the same temporal space, can be used as a measure of their evolutionary divergence. The respective probability distributions of the temporally overlapping pairs of SARS-CoV-2 variants are shown in Figure 7.8 (a) to (f). This aspect will be further discussed and elaborated for the evaluation of the Indices of Divergence of the variants.

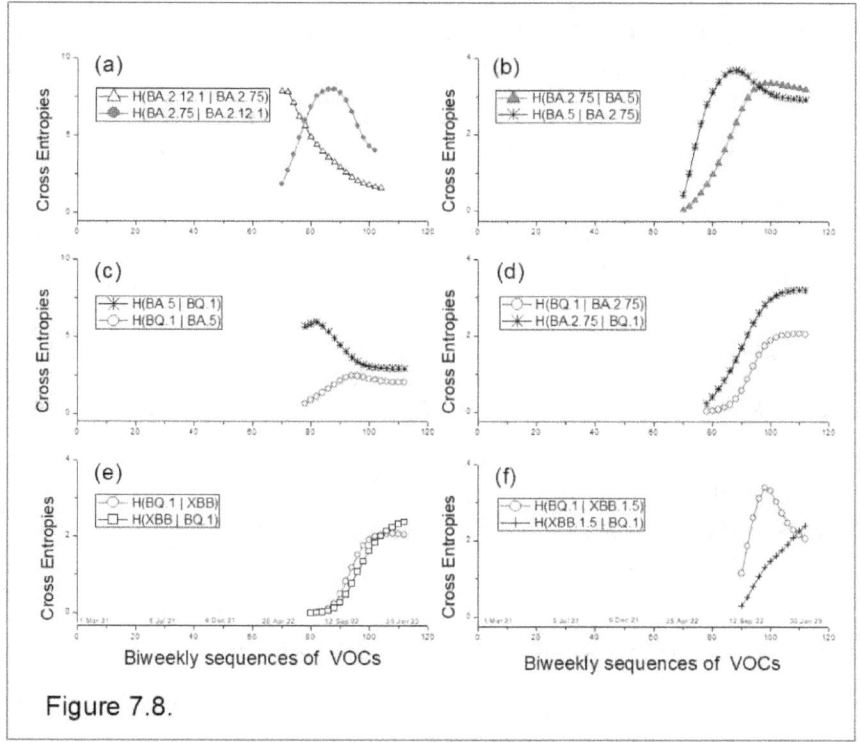

Figure 7.8.

Figure 7.8. The cross entropies are plotted for the six combinations of the Omicron variants. (a) BA.2.12.1 & BA.2.75. (b) BA.2.75 & BA.5. (c) BA.5 & BQ.1. (d) BA.2.75 & BQ.1, (e) BQ.1 & XBB. (f) BQ.1 & XBB.1.5.

Figure 7.8. has the cross entropic plots for the six combinations of the Omicron variants that are chosen due to their temporal overlap with each other. The objective of the exercise is to understand their mutual relationship vis-à-vis their cross-emergence profiles. Figure 7.8(a) is plotted for the cross entropies resulting from the overlap of the probability distributions of BA.2.12.1 & BA.2.75. Similarly, 7.8(b) is for the pair BA.2.75 & BA.5. Figure 7.8(c) is for BA.5 & BQ.1, (d) has been plotted for BA.2.75 & BQ.1, (e) for the distributions of BQ.1 & XBB and finally, (f) represents the twin cross entropies of the probabilities of the variants BQ.1 & XBB.1.5.

TABLE 7.4.   CROSS ENTROPIES, CROSS EMERGENCE FUNCTIONS, AND INDICES OF DIVERGENCE

VOCS PAIRS	$H(p|q)$	$H(q|p)$	$\varepsilon(p|q)$	$\varepsilon(q|p)$	$\Delta\varepsilon$
DELTA & BA.1	39	20.57	1232.4	249.3	983
BA.1 & BA.2	20.46	23	248	361	113
BA.2 & BA.2.12.1	4.9	3.7	76.93	57.87	19
BA.2.12.1 & BA.5	16.3	25.19	255	633.3	378
BA.5 & BQ.1	6-3	0.6 - 2	75	31	~82
BA.2.75 & BQ.1	3.18	2	66.14	31	35
BQ.1 & XBB	2.06	2.4	31.9	35.3	3.4
BQ.1 & XBB.1.5	2.2	2.2	34	18	16
BA.2.12.1 & BQ.1	5.8	5.8	91	90	~0
BA.2.75 & BA.5	3.28	3.28	70.7	93	22.3

Table 7.4 has been constructed using the data presented in Figures 7.3 to 7.8 for the cross entropies, Cross Emergence Functions, and the resulting Indices of Divergence between the chosen pairs of the Omicron variants. The first pair of the variants includes the Delta and BA.1. Delta characterized as the most significant Early VOC was replaced by the first of the Omicrons- BA.1.

7.8. Summary of Emergence and Cross Emergence of the variants of SARS-CoV-2 in USA.

This chapter is dedicated to the study of, and the investigations of the emergent behavior, and the profiles of cross emergence of the variants of SARS-CoV-2 in the United States. The analysis presented here is based on the 4 million sequences conducted during the 112 weeks starting on the 1st of March 2021. The data, as reported by ref. [85,86] was shown in Figures 7.1 and 7.2. In the set of Figures 7.3 to 7.8, the emerging and cross emerging patterns were presented and analyzed. The characteristic features of the evolving trends of SARS-CoV-2 from the two years' data of the biweekly sequences of the variants were tabulated as the set from Tables 1 to 4. Here, the emergence (λs), and the cross-emergence profiles through the CEFs and the calculated Indices of Divergence $\Delta\varepsilon$ amongst the variants of the evolving SARS-CoV-2 are summarized.

1. The Emergence of variants of SARS-CoV-2. Emergence Function λ was used as an instrument of comparing the emerging traits of the successive stages of the pandemic in Chapters 4, 5 and 6. The present chapter has presented the successive series of the emergence of the distinct lineages or a sub-lineages from an existing lineage, in the earlier sections of the present chapter in Figures 7.3 to 7.8 and tabulated as Table 1.

A comprehensive visualization of evolution of the variants for the 2-years' data analyzed in the previous sections 7.1 to 7.7, is presented in Figure 7.9. The figure is based on the temporal evolutionary profiles through the instantaneous entropic graphs and the corresponding λs for each variant. Figure 7.9 identifies the distinct phases of the emergence of SARS-CoV-2 variants starting from the Pre-VOCs to the XBB variants of the Omicron family. In Figure 7.9, the calculated Emergence Functions are displayed as bar graphs for the successive variants. Figure 7.9 is the collection of the Emergence Function λ for each of the VOCs shown earlier in Figure 7.5(a).

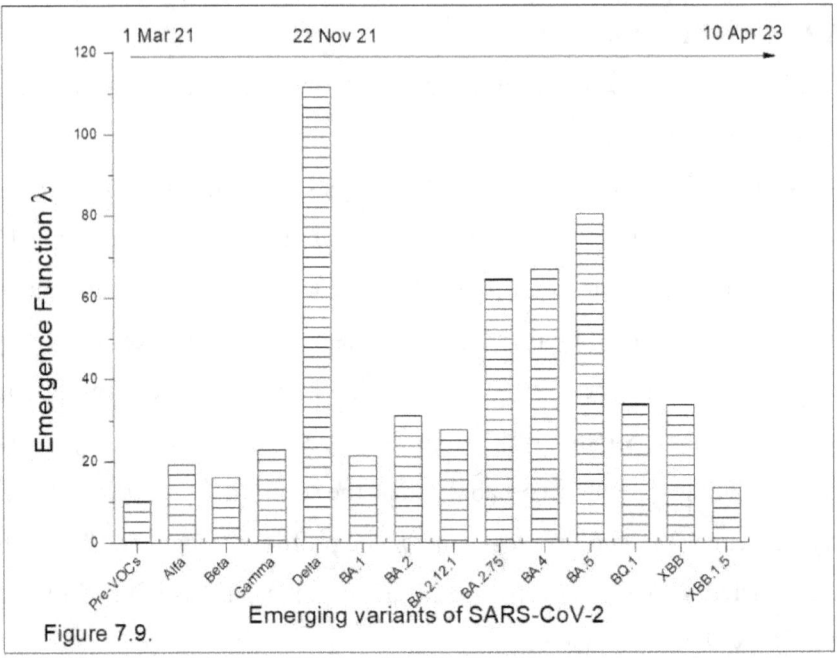

Figure 7.9. Emerging variants of SARS-CoV-2

The bar-graphs for of the Emergence Function λ of the series of variants illustrate:

a. The temporal signatures of the emerging variants.

b. The unique place for each VOC or the sub-lineage, on the tree of Emergence of SARS-CoV-2 variants.

c. The dynamics of the structural and functional diversification of the variants reveals the changing fitness parameters.

d. The consistency of the diversifications of the variants seems to be based on the selective processes induced by the ambient conditions i.e., social distancing, lockdowns, vaccinations etc.

e. The Emergence functional parametrization of the relative selective strength of the mutants in terms of the higher acquired transmissibility characteristics leading to the new variants.

2. The Cross Emergence Functions. The temporal co-existence of the pairs of variants represented by their respective probability distributions provides the basic condition for evaluating the CEFs. The cross entropies are evaluated for the new, emerging variants, represented by the probability distribution function p, and the pre-existing, declining variants with the corresponding distribution q. For the phase 1 where the Pre-VOCs and the Early VOCs were analyzed, the respective CEFs denoted as $\mathcal{E}(p|q)$ and $\mathcal{E}(q|p)$ for the chosen pairs of VOCs were calculated from the twin graphs for the cross entropies of the existing lineage and the emerging ones in Figures 7.3 to 7.8 and tabulated in Tables 7.1 and 7.2. Figure 7.10(a) has the CEFs from Table 7.1 plotted as vertical bars between the Pre-VOCs and the set of four Early VOCs Alfa, Beta, Gamma, and Delta.

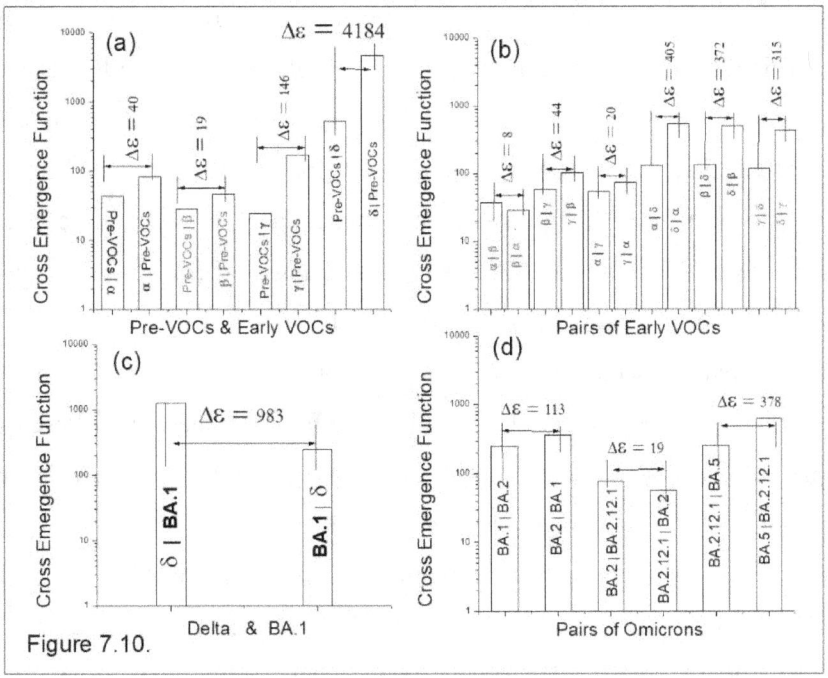

Figure 7.10.

Figure 7.10(b) shows the mutually evaluated CEFs of the six pairs of the

Early VOCs. The Indices of Divergence are shown for each pair. The numerical values for the CEFs evaluated for the VOCs with Pre-VOCs follow Gamma ($\mathcal{E}(p|q)$=24.3), Beta ($\mathcal{E}(p|q)$=28.41), Alfa ($\mathcal{E}(p|q)$=43.1), and Delta ($\mathcal{E}(p|q)$=522.7). The corresponding CEFs ($\mathcal{E}(q|p)$) of the Pre-VOCs with the VOCs increase moderately in magnitude (($\mathcal{E}(q|p)$) > ($\mathcal{E}(p|q)$)) in the case of Beta (($\mathcal{E}(q|p)$=47.1), Alfa (($\mathcal{E}(q|p)$=83.38), and Gamma (($\mathcal{E}(q|p)$=171). The CEF is exceptionally large in the case of Delta ($\mathcal{E}(q|p)$=4707).

3. The corresponding Indices of Divergence $\Delta\mathcal{E}$ are shown for each combination in Figure 10(a) and (b). The Index of Divergence $\Delta\mathcal{E}$ is treated as the mutual dynamic measure of the informational divergence between any two temporally coincident mutants. The transition from the Early VOCs to the Omicrons depicted through the graphs for the instantaneous entropic emergence of the Early VOCs and the Omicrons shown above in Figure 7.5(a) is indicated by the relative magnitudes of the respective Emergence Functions (λs) in Figure 7.9. Delta has the largest Emergence Function λ.

The BA.1 Omicron that emerged, coexisted for a while, and then replaced Delta, is shown with comparatively smaller λ. The Cross Emergence Functions of the Delta and BA.1 is plotted in Figure 7.10(c). The Index of Emergence $\Delta\mathcal{E}$=983 for the pair [Delta | BA.1].

Three pairs are selected from Table 7.4 and plotted in Figure 7.10(d) out of the sequence of the VOCs from Delta to XBB.1.5. These are chosen to represent the magnitudes of the Indices of Divergence of the evolving Omicrons, BA.1, BA.2, BA.2.12.1, and BA.5.

Amongst the temporally overlapping variants in Figure 7.10(c) and (d). Delta and BA.1 have the largest $\Delta\mathcal{E}$ (=983), followed later by the pair of variants BA.1 and BA.2 with $\Delta\mathcal{E}$ =113. The evolution of BA.5 from BA.2.12.1 is the next prominent transition with $\Delta\mathcal{E}$ =378.

The important aspects of the emerging variants are displayed through the

Indices of Emergence $\Delta\mathcal{E}$ amongst the chosen pairs of VOCs:

a. Temporal divergence in the case of Delta and BA.1 implied the sustenance of Delta Variant for the longest periods, the longer existence in terms of ζ_0, and exhibiting the largest IoD $\Delta\mathcal{E}$.

b. The Delta-replacing variant the BA.1 displayed higher transmissibility, improved immune escape to be able to spread in the increasingly restrictive environment.

c. After the next few mutations, the variant BA.5 emerged with a relatively superior survival kit of higher transmission and immune escape.

d. The later Omicrons consistently displayed relatively smaller magnitudes of $\Delta\mathcal{E}$. This may be the sign of the end-approaching evolutionary sequence of the SARS-CoV-2.

The Graphical Summary presents the information-theoretic, quantitative description of the evolution of the variants of SARS-CoV-2 in the USA during ~ 115 weeks starting from 1 Mar 2021, through the Cross Emergence Functions and the associated Indices of Divergence $\Delta\mathcal{E}$ amongst the neighboring VOCs present the relative Cross Emergent characteristics of the evolving ones with respect to the neighbor/parent/sister variants.

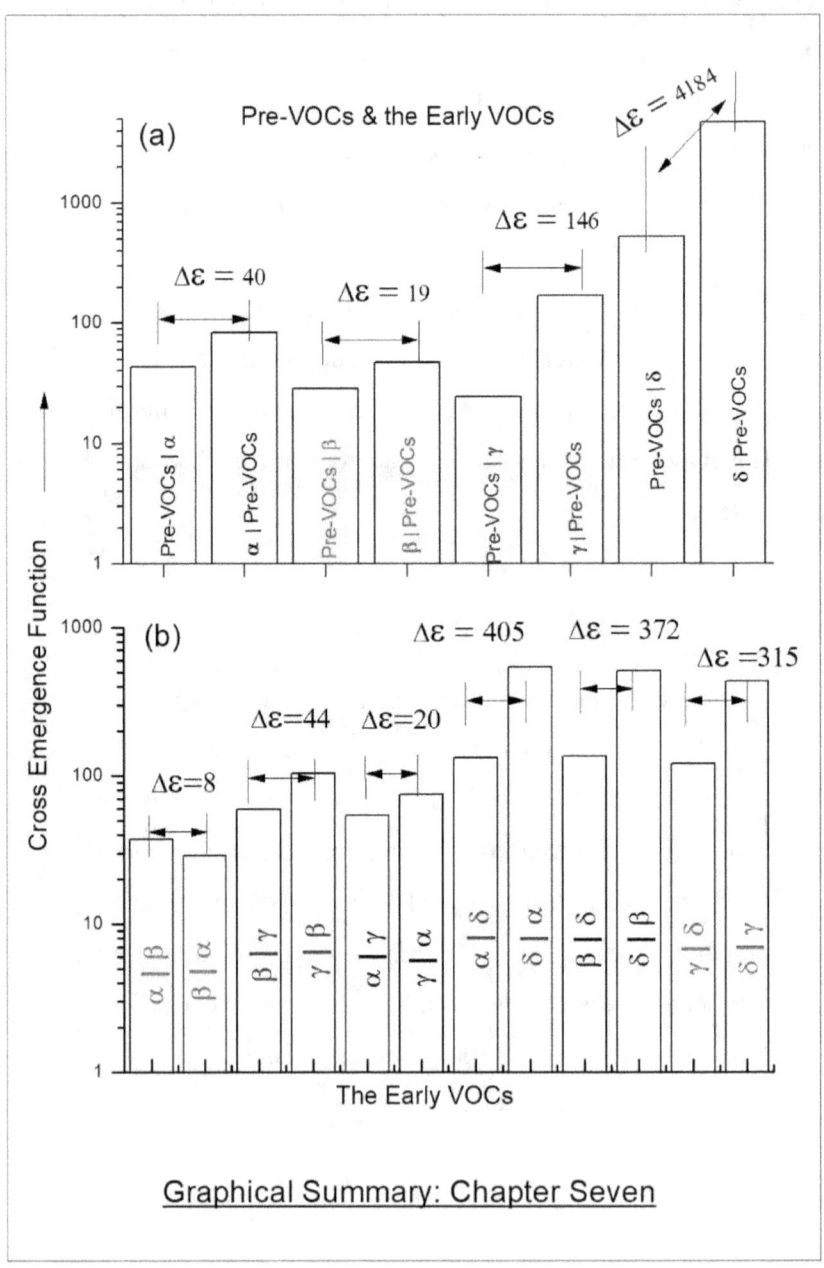

Graphical Summary: Chapter Seven

CHAPTER 8

SELF-ORGANIZING, EMERGING, CROSS-EMERGING SARS-CoV-2

8.1. The Diversity of Emergence. In the last five chapters, the severe acute respiratory syndrome coronavirus-2 (SARS-CoV-2), that emerged in the form of the two distinct episodes in 2002 and 2019, has been analyzed as self-organizing, information-generating and manipulating dynamical system. The information-theoretic emergence model developed in the first two Chapters for the pandemic, was based on constructing (a) the cumulative probability distribution functions for the cases during the individual stages/waves and (b) for the population of the cases during the whole duration of the pandemic. The Emergence Functions (λs) were calculated by using these pandemic-defining probability distributions. In Chapter Three, the Emergence Functions for the total duration of the first episode of SARS-CoV-2, were evaluated, presented, and discussed with emphasis placed on China and Hong Kong. The recent episode was analyzed in Chapters 4 to 6. The λs for the six(6) WHO-regions were plotted as a function of the weeks of COVID-19, in Figure 4.8 of Chapter Four. Similarly, the λs for the case study of the five chosen countries of the three WHO-regions Europe, W Pacific, and the Americas, were presented for the first 100 days and the 166 weeks of COVID-19, in Figures 6.7 and 6.8 of Chapter Six. The emerging stages/waves and phases of the pandemic was demonstrated through the Emergence Functions, the Cross Emergence Functions, and the Indices of Divergence in the previous Chapters.

In Chapter 6, the emergence and cross emergence of the pandemic in the three regions were investigated by utilizing the data on the COVID-19 infected cases, USA being one of the five chosen countries. In Chapter 7, pandemic in the USA, was exclusively studied by utilizing the data on the biweekly sequencing of the infected patients to investigate the emergence of the variants of SARS-CoV-2. The evolving trends of SARS-CoV-2 from the two years' data of the biweekly sequences of the variants were calculated, tabulated as the sets of Emergence Functions in Tables 7.1 to 7.4, and plotted as bar graphs for each variant in Figure 7.9 of Chapter Seven. The evolution of the variants of SARS-CoV-2 was demonstrated through the diversity of their respective λs. The phenomenon of Emergence displayed the built-in pathways for diversity as the essential requirement for the enhanced transmissibility of the virus.

The information-theoretic model developed and applied to understand the self-organizing, information-manipulating character of the pandemics provide clues about the nature of the Open-ended emergence. This will help us conclude the Emergence and Cross Emergence of COVID-19 with the cases almost approaching ~800 M in 2024 and continuing to increase, albeit at relatively low levels.

8.2. Selectivity amongst the Diversified Emergence. In Chapters 3 to 6, the cross-emergence analyses were performed, exclusively for the multiple stages/waves of the infected cases. The waves of the cases were characterized by the Cross Emergence Functions (CEFs) $\mathcal{E}(p|q)$ and $\mathcal{E}(q|p)$ between the variants with probability distributions $p(\zeta)$ and $q(\zeta)$. However, the distinctions of the cases of the respective stages/waves based on the identification/sequences of the variants of SARS-CoV-2, was delayed until Chapter Seven. It was shown in Chapter Six that the analyses based on the probability densities constructed from the number densities of the cases, can help characterize the emergence of the successive stages/waves of the pandemic at the local levels for the chosen countries, and for the WHO-designated regions, and for the global pandemic. The Indices of Divergence

ΔƐ provided the measures of distinctiveness as well as the divergence amongst the waves/stages. In Chapter Seven, the same Emergence and Cross Emergence model provided the tools λs and ΔƐs to understand and describe the evolutionary profiles of mutations of the variants as was clearly demonstrated in the Graphical Summary of Chapter Seven.

8.3. Scale invariance and sensitivity to initial conditions of the pandemic. The emergence of the two episodes of the coronavirus pandemics, separated by seventeen (17) years, display the fractal-like characteristics with certain exceptions. SARS-CoV-2 spread to ~10^4 persons in 2002-4, mostly in 29 countries of the W Pacific region. Starting in late 2019, and still ongoing pandemic due to the same virus, has infected ~10^9 persons globally. This clearly implies the difference of five orders of magnitude in the number of cases. The pandemic demonstrates scale-invariance. The dynamical systems display fractal-like characteristics through scale-invariance. Here, the pandemic, during the two temporally disconnected episodes, under diverse conditions of emergence, has displayed scale invariance. COVID-19 displayed fractal character at the local levels (cities, countries), and at global scale (regions and continents) through scale invariance of the number of cases as well as projecting self-similarity of the emerging, cumulative ensembles of the infected populations even though these vary in numbers and their respective Emergence Functions.

The previous five Chapters have provided the profiles of the two pandemics' emergence that clearly display sensitivity to initial conditions. Here, in the context of the spreading and mutations of the virus, the initial conditions can be defined by investigating the profiles of the cross emergence under varied conditions. The Index of Divergence ΔƐ provides the quantitative assessment of the diversification through mutations. It has been treated as the dynamic measure of divergence between different waves of the cases and the interacting mutatnts.

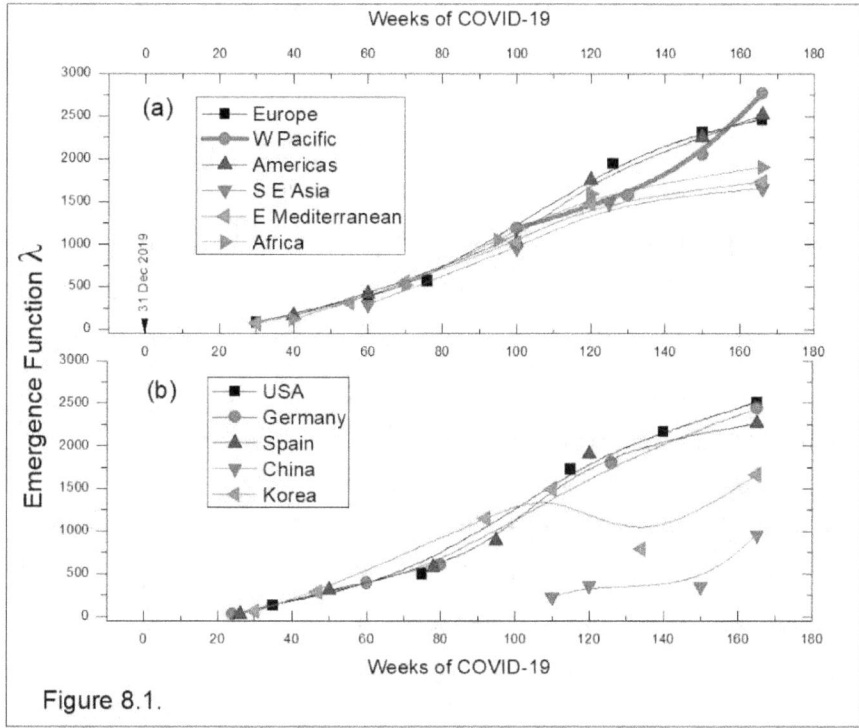

Figure 8.1.

Figure 8.1. The scale-invariance and self-similarity of the emerging pandemic can be observed in the Emergence Function versus the weeks of COVID-19. (a) In the six (6) WHO-designated regions Europe, W Pacific, Americas, S E Asia, E Mediterranean, and Africa, the evaluated λs for successive stages/waves of the cases during the 166 weeks of COVID-19 show similar profiles; except the W Pacific region which shows marked differences in its λ(W Pacific). (b) The λs are plotted for the five chosen countries from three regions Europe (Germany and Spain), W Pacific (China and Korea) and the USA from the Americas. Again, the λ-profiles for the two w Pacific region do not match the corresponding λs for the other three countries.

There are significant differences between the two episodes of the pandemic not only in the number of infected cases, but also in the mutation rates and the evolution of the variants. As opposed to the emergence of the earlier episode in 2002-4, in the case of COVID-19, the evolutionary trail has identified substantial number of variants. This can be partially explained by the sensitivity to initial conditions. However, in this respect, the distinction is clear between the two episodes of the SARS-CoV-2 pandemics; the present scenario is the evolution of the virus that has been meticulously probed, investigated and reported.

The nature of the conditions and constraints in diverse environments have been shown to strongly dictate the cross-emergence of the pandemic. Different social and cultural environments, levels of hygiene, the state-imposed and voluntary restrictions on public gatherings, quarantines and vaccinations impact the probabilities of infections. Mutations favoring enhanced transmissibility under the restrictive conditions have been reported to emerge from their sister lineages/variants.

The information-theoretic analysis of the Emergence and Cross Emergence of SARS-CoV-2 in the USA during the two years from 1 Mar 2021 to 10 April 2023 are summarized below by using the tools developed and employed in this book.

1. Starting with the initial, few cases due to the 'original virus,' optimum number of the infected cases are produced. From few to many became the rule for the emergence of pandemic.
2. Evolution of the new and novel variants have enhanced the transmissibility and higher chances of survival in the diverse and varied environments.
3. Diversity has been demonstrated in the detailed analyses presented in Chapters 3 to 6, as the rule by which the pandemic spreads from the initially exposed few individuals to large communities, at different rates of transmission, which are dictated by the diversity-inducing environments of the targeted populations.
4. Diversity also implies the emergence of multiple variants that allow the virus to adjust according to the diversified environments. The continued

transmission through resistant sets of population induce mutations that allow higher levels of survival and transmission. The dominant variant generally acquire and display the largest Emergence Function λ, like the Delta VOC, and the Omicrons BA.1 and BA.5.

5. Cross Emergence Functions (CEFs) of the temporally coincident variants, have been investigated in Chapter 7. These show marked divergences for the dominant, longer lasting variants in their respective CEFs.

6. The study of the emerging profiles of the CEFs of the sub lineages and the variants that share the same temporal space, but different physical locations demonstrate their relative effectiveness of transmissibility, immune escape etc.

In Chapters 3 to 6, the Indices of Divergence $\Delta\mathcal{E}$ have been employed in studying the relative comparisons of the emerging trends displayed by the stages/waves of the pandemic at various levels, local, regional and the global. In Chapter 7, $\Delta\mathcal{E}$ has been extensively used to identify the dominant emergent variants.

The case of the W Pacific's divergence from the global pandemic is amply displayed in Figure 8.2(a) and (b). The pandemic after starting in China, spread to the neighbors, but within the first 100 days its Chinese profile diverged substantially from the trends of the increasing cases elsewhere as the bar graphs show in 8.2(a). Therefore, the Chinese and the global cross emergent features display large divergence via $\Delta\mathcal{E}$.

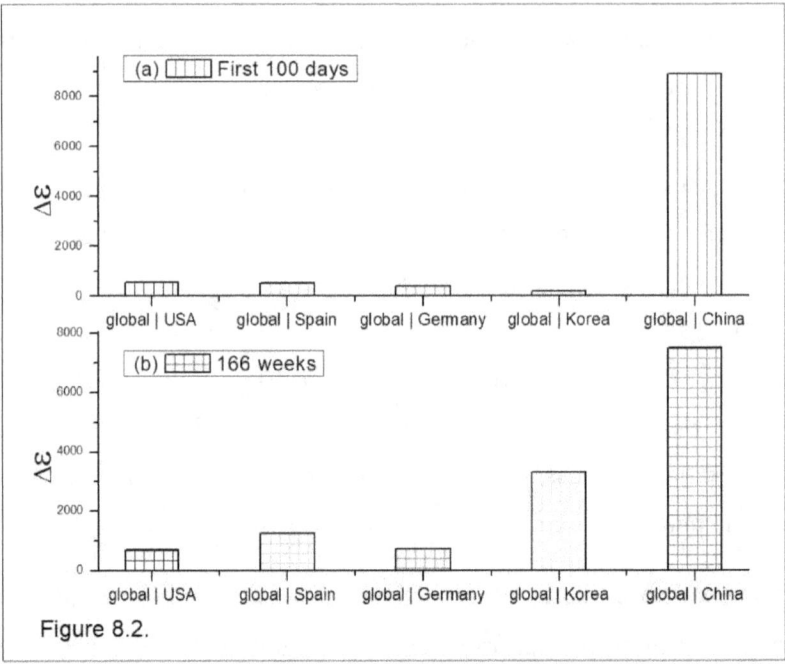

Figure 8.2. Scale invariance and self-similarity of pandemic's Indices of Divergence between the global and the five countries of the three regions is clarity demonstrated in the bar graph plots in (a) for the duration of the first 100 days and in (b) during the entire duration of the 166 weeks.

The two W Pacific countries showed divergent trends of the emergence of the pandemic throughout the 166 weeks, as compared with the three countries Germany, Spain, and the USA. The regions of Europe and Americas seem to have been the fertile ground for the spread of the pandemic; the two share ~59.2% of the total COVID-19 cases. Therefore, the divergence between the two regions and the global pandemic is minimal as shown in the graphs for the respective Indices of Divergence $\Delta \mathcal{E}$ of Figure 8.2(a) and (b).

8.3. The Emergence of SARS-CoV-2 variants in the USA. Figure 8.3 was presented in Chapter 7 as Figure 7.9, to display the calculated

Emergence Function λ of the mutating Variants of Concern (VOCs) of SARS

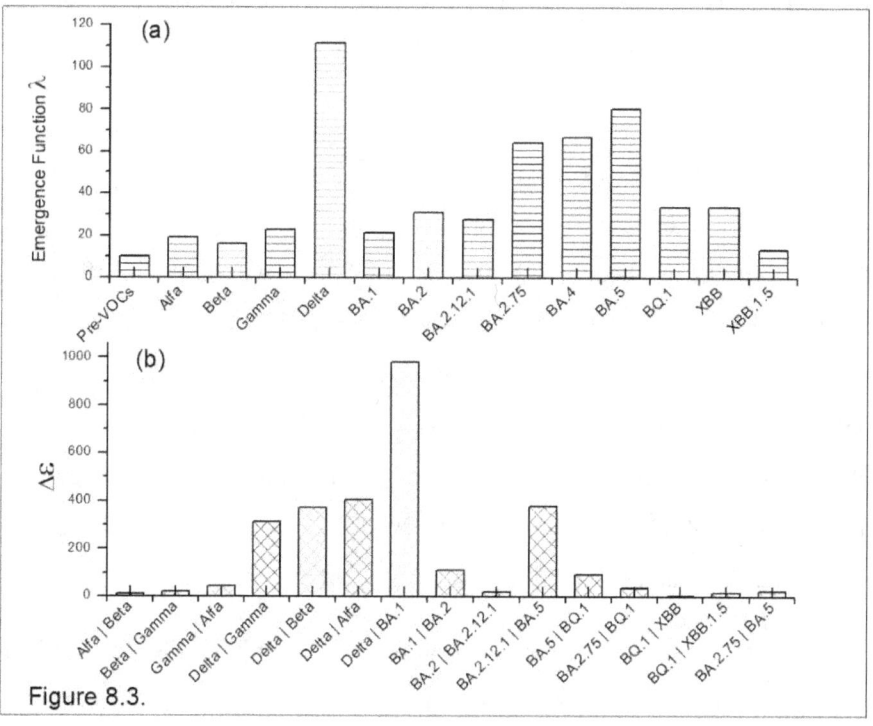

Figure 8.3. CoV-2.

Figure 8.3. (a) The Emergence Funtions and (b) the Indices of Divergence ΔƐ for the emerging Variants of Concern of SARS-CoV-2 are shown sequentially, as these emerged, flourished and became extinct over the 110 weeks starting from 1 March 2021.

Here, in Figure 8.3, these are presented again as the footprints of the evolution of SARS-CoV-2 based on the data of the infected cases that was used to evaluate λ. The data was obtained from the biweekly sequencing of the VOCs in the USA between 1 March 2021 and 10 April 2023. The first of the variants sequenced was the Pre-VOCs that had dominated before the emergence of the Early VOCs. Similarly, the figure showed the continuing growth, beyond April 2023, of the number of the cases due to the Omicron variants BQ.1, XBB and XBB.1.5. These latest Variants were still emerging to their full capacity and may lead to other variants. Hence, their λs appear as the under-represented graphs in Figure 8.3(a). All other variants are fully represented in the figure, implying that their temporal history of the initiation, spread and extinction was recorded within the two years of the data acquisition.

The Emergence profile of the VOCs tabulated through Emergence Function λ in Figure 8.3(a) provides clues about the emergent trends and the direction of the pandemic: (a) A series of VOCs with a range of low λ emerged during the transition from a dominant variant like Pre-VOCs until the Delta VOC emerged and dominated all others. Pre-VOCs dominated before the emergence of the Early VOCs that included Alfa, Beta, Gamma, Delta, and a few others. (b) The peak of the pandemic existed between and around the VOCs with large λ, From Pre-VOCs $\rightarrow$ Delta. The Omicrons share gradually rising λ towards BA.5, followed by a decline.

The Cross Emergence Function is depicted in Figure 8.3(b) through the Indices of Divergence $\Delta\mathcal{E}$ that complement the above description of the dominant variants and their role in the spreading of the pandemic. The emerging variants are displayed temporal divergence as shown in the case of Delta and BA.1 with Deltaexhibiting the largest IoD $\Delta\mathcal{E}$. The Delta-replacing variant the BA.1 displayed higher transmissibility, had improved immune escape and was able to spread in the increasingly restrictive environment. Similarly, after the next few mutations, the variant BA.5 emerged with relatively superior survival kit of higher transmission and immune escape. The later Omicrons can be seen in Figure 8.3(b) consistently displaying smaller magnitudes of $\Delta\mathcal{E}$. Does it indicate the end-approaching evolutionary sequence of the SARS-CoV-2 or a symbol of the Open-ended evolution of SARS-CoV-2?

8.4. The Four Paradigms of Evolution of SARS-CoV-2. The four consequent paradigms of the virus' evolutionary trail can be introduced.

These describe the emergent, self-organizational framework of the operative stages of the evolution of SARS-CoV-2. In addition to explaining the evolutionary traits of SARS-CoV-2, these seem to provide clues to the Open-ended Emergence of the pandemic.

1. The first paradigm deals with the optimal production of the Virus, demonstrated to be the result of cooperative phenomena dictated by the initial conditions of the active ingredients and the environment. Optimum production establishes the conditions for the emergence of the virus that may lead, later, to mutations during the spreading of the pandemic. The Emergence Functions λ of each of the initial stages of COVID-19, provided the relative measures of the temporal dimensions and the information generated by the expanding, infected populations of the cases, in distinct places and regions. Emergence of the mutating variants, sharing the same temporal space, is also defined through their Emergence Functions. The variants' emergence in various social and cultural environments is filtered by the protective methods, remedial practices adopted by the public at large, and the virus-avoiding measures that included social distancing, lockdowns, personal hygiene, and vaccinations. Emergence Functions describing the distinct, initial stages, can be used to quantify the local epidemic evolving into the full-blown pandemic.

2. The second paradigm deals with the multiplicity of the emerging and accumulating, dynamic ensembles of the variants. The diversity is based on intense, competitive activities amongst the evolving variants, and with their respective environments. The factors and forces responsible for the diversity of ensembles of the variants are demonstrated in terms of their cumulative $\{\sum_x \lambda_x (= \zeta_0 * H_m)\}$ as well as the individual Emergence Functions (λs). Selection of the dominant, longer lasting variant displays the largest λ.

3. The third paradigm is based on the Cross-Emergent measures to quantify the selective criteria of the evolution of the variants. The relative magnitudes of the Cross Emergence Functions $\mathcal{E}(p|q)$ and $\mathcal{E}(q|p)$ between the two coexisting variants with probability distributions $p(\zeta)$ and $q(\zeta)$, indicate and identify the dominant, and the longer lasting species. The dominant variants with large Emergence Functions also display the larger Cross Emergence Functions as compared with the other variants that share the same temporal space. This leads to the selective advantage displayed through the higher values of the respective CEFs, in favor of the dominant variants, that had demonstrated larger λs.

4. The fourth paradigm derived from the evolution of SARS-CoV-2, relates to the later stages of the pandemic where the sequential emergence of multiple variants start to occur with approximately similar orders of magnitude CEFs. These imply that the IoDs ($\Delta\mathcal{E}$) would be decreasingly small for later variants. At this stage, the dynamical system of the pandemic seems to have settled towards the final stages of evolution through the selection of the mutating variants that keep on increasing their transmissibility, yet do not show large Indices of Divergence $\Delta\mathcal{E}$. These could be regarded as the stable states of the Open-ended Cross Emergence.

The Graphical Summary. The description of the sequential route of the self-organizing, emerging, and cross emerging dynamical system of COVID-19 is described. Starting with the optimum productivity of the virus (in terms of the numbers of the infected cases) → the diversity of Emergence; implying that the infected population becoming the carriers of the original variant and the multiple emerging variants of the Virus → the Cross Emergence that further lead to the selection of the variant(s) with the relatively higher possibilities of survival and transmissibility, expressed through the respective Indices of Divergence (IoD) $\Delta\mathcal{E}$. The IoDs not only identify the dominant species, but also provide the quantitative measure of the divergence between the emerging, dominating variants and the diminishing ones.

Self-Organizing Emerging, Cross Emerging COVID-19

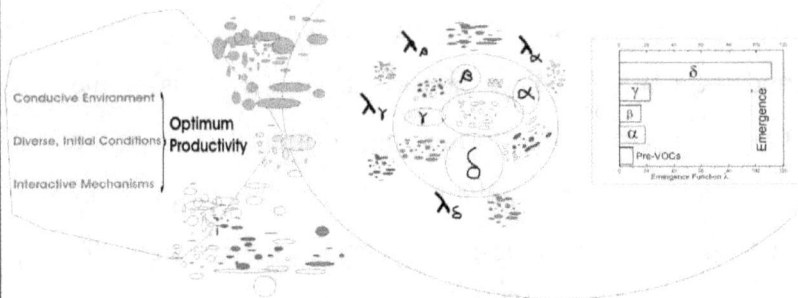

Emerging Ellipse of Diversity

Cross Emergence among the Variants & Environment

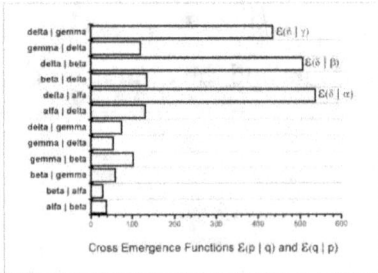

The Selected Variant(s)

Graphical Summary: Chapter Eight

Optimum productivity — Diversity of Emergence — Cross Emergence — Selectivity based on Index of Divergence

List of References

1. W. O. Kermack and A. G. McKendrick, "A contribution to the mathematical theory of epidemics." Proc. Roy. Soc. A **115**, 700 (1927).
2. A. Vazquez, "Epidemic outbreaks on structured populations," J Theor Biol, **245**, 125 (2007).
3. B. Kotnis and J. Kuri, "Stochastic analysis of epidemics on adaptive time varying networks," Phys. Rev. E. Stat. Nonlin. Soft Matter Phys. **87**, 062810, (2013).
4. D. Clancy and P.D. O'Neill, "Bayesian estimation of the basic reproduction number in stochastic epidemic models," Bayesian Anal. **3**, 737 (2008).
5. J. Dureau, K. Kalogeropoulos and M. Baguelin, "Capturing the time-varying drivers of an epidemic using stochastic dynamical systems," Biostatistics **14**, 541 (2013).
6. A. G. Hoen, et.al., "Epidemic Wave Dynamics Attributable to Urban Community Structure: A Theoretical Characterization of Disease Transmission in a Large Network," J. Med. Internet Res, **17**(7), (2015). PMID: 26156032.
7. L. J. S. Allen, "A primer on stochastic epidemic models: Formulation, numerical simulation, and analysis," Infectious Disease Modelling **2**, 128 (2017).
8. J. Tolles and T. Luong, "Modeling epidemics with compartmental models," JAMA **323**, 2515 (2020).
9. O. N. Bjørnstad, K. Shea, M. Krzywinski and N. Altman, "Modeling infectious epidemics", Nat Methods. **17**, 455 (2020).
10. J. Meagher and N. Friel, "Assessing epidemic curves for evidence of superspreading," J R Stat Soc Ser A Stat Soc. **185**, 2179 (2022).
11. G. Marion, et al., "Modelling: Understanding pandemics and how to

control them" Epidemics, **39**, 100588 (2022).

12. J. Enright and R. R. Kao, "Epidemics on dynamic networks," Epidemics, **24**, 88 (2018).

13. A. J. Kucharski et al., "Early dynamics of transmission and control of COVID-19: A mathematical modelling study", Lancet Infect. Dis. **20**, 553 (2020).

14. A. Melnyk, L. Kozarov and S. Wachsmann-Hogiu, "A deconvolution approach to modelling surges in COVID-19 cases and deaths", Scientific Reports, **13**, 2361 (2023).

15. M. A. Khan and A. Atangana, "Mathematical modeling and analysis of COVID-19: A study of new variant Omicron", Physica A **599**, 127452 (2022).

16. M. Chinazzi, et al., "The effect of travel restrictions on the spread of the 2019 novel coronavirus (COVID-19) outbreak", Science **368**, 395 (2020).

17. A. L. Bertozzi, et.al., "The challenges of modeling and forecasting the spread of COVID-19", PNAS, **119**, 16732 (2020).

18. S. Riley et al., "Transmission dynamics of the etiological agent of SARS in Hong Kong: Impact of public health interventions," Science **300**, 1961 (2003).

19. J. Wallinga, P. Teunis, "Different epidemic curves for severe acute respiratory syndrome reveal similar impacts of control measures," Am. J. Epidemiol. **160**, 509 (2004).

20. S. Cauchemez et al., "Real-time estimates in early detection of SARS," Emerg. Infect. Dis. **12**, 110 (2006).

21. A. J. Lotka, "Analytical Note on Certain Rhythmic Relations in Organic Systems", Proc. Nat. Acad. Sci. **6**, 410 (1920).

22. R. Pearl and R. J. Read, "On the Rate of Growth of the Population of the United States Since 1790 and its Mathematical

Representation" Proc. Nat. Acad. Sci. **6**, 275 (1920).

23. R. M. May, "Biological Populations with Nonoverlapping Generations: Stable Points, Stable Cycles, and Chaos," **186**, 645 (1974).

24. E. Renshaw, "Modelling Biological Populations in Space and Time," Cambridge University Press, (1991).

25. R. M. May, "Simple Mathematical Models With Very Complicated Dynamics," Nature, **261**, 459 (1976).

26. R. M. May, *Stability and Complexity in Model Ecosystems*, Princeton University Press, Princeton (1973).

27. R.L. Devaney, *A First Course in Chaotic Dynamical Systems: Theory and Experiment*, Addison-Wesley (1992).

28. S. Strogatz, *Nonlinear Dynamics and Chaos: With Applications to Physics, Biology, Chemistry, and Engineering*, Addison-Wesley (1994).

29. R. Holmgren, *A First Course in Discrete Dynamical Systems*, 2nd edition, Springer-Verlag (1996).

30. D. K. Arrowsmith, L. M. Place, *Dynamical Systems: Differential equations, Maps, and Chaotic Behaviour*, Chapman and Hall/CRC (1992).

31. K. Alligood, T. Sauer & J. Yorke, *Chaos: An Introduction to Dynamical Systems*, Springer (2000).

32. M. W. Hirsch, S. Smale, *Differential Equations, Dynamical Systems and Linear Algebra*, Academic Press, London (1974).

33. L. Perko, *Differential Equations and Dynamical Systems*, Springer-Verlag (1991).

34. R. C. Robinson, *An Introduction to Dynamical Systems: Continuous and Discrete*, Prentice Hall (2004).

35. Y. Bar-Yam, *Dynamics of Complex Systems*, Addison-Wesley,

Reading (1997).

36. G. C. Layek, *An Introduction to Dynamical Systems and Chaos,* Ch 1,2,4, Springer, New Delhi (2015).
37. S. Wiggins, *Introduction to Applied Nonlinear Dynamical Systems and Chaos*, 2nd Ed. Springer (2003).
38. R. C. Hilborn, *Chaos and Nonlinear Dynamics: An introduction for Scientists and Engineers,* Oxford University Press, Oxford (2000).
39. C. E. Shannon, "A Mathematical Theory of Communication," The Bell System Technical J. **27**, 379 and 623 (1948).
40. C. E. Shannon and W. Weaver, *The Mathematical Theory of Communication*, University of Illinois Press (1998).
41. S. Ahmad, "Information generating, sharing and manipulating Source-Reservoir-Sink model of self-organizing dissipative structures," Chaos **28**, 123125 (2018).
42. R. Shaw, *The Dripping Faucet as a Model Chaotic System*, Aerial Press, Santa Cruz, California (1984).
43. H. Haken, *The Science of Structure: Synergetics*, Van Nostrand Reinhold, New York (1981).
44. H. Haken, "Cooperative phenomena in systems far from thermal equilibrium and in nonphysical systems," Rev. Mod. Phys. **47**, 67 (1975).
45. I. Prigogine and I. Stengers, *Order Out of Chaos, Man's New Dialogue with Nature,* Bantam Books, New York (1984).
46. G. Nicolis and I. Prigogine, *Self-organization in non-equilibrium systems* (John-Wiley & Sons, New York (1977).
47. M. M. Waldrop, *Complexity: The emerging science at the edge of order and chaos*, Touchstone Simon & Schuster, New York, (1992).
48. B. B. Mandelbrot, *The Fractal Geometry of Nature* (W. H. Freeman & Co., New York (1977).

49. M. Barnsley, *Fractals Everywhere*, 2nd edition, Academic Press (1993).

50. B. Davies, *Exploring Chaos: Theory and Experiment*, Westview Press (2004).

51. P. Berge, Y. Pomeau and C. Vidal, *Order within Chaos,* John Wiley & Sons, New York, (1984).

52. P. Corning, "The Re-Emergence of "Emergence": A Venerable Concept in Search of a Theory," Complexity, **7**, 18 (2002).

53. J. Holland, *Emergence: from Chaos to Order*, Addison Wesley, Reading (1998).

54. D. Blitz, *Emergent Evolution: Qualitative Novelty and the Levels of Reality*, Kluwer Academic Publishers, Dordrecht (1992).

55. J. D. Halley and D. A. Winkler, "Consistent concepts of self-organization and self-assembly," Complexity **14**, 10 (2008).

56. J. Skar, "Introduction: Self-Organization as an Actual Theme," Phil. Trans. R. Soc. Lond. A **361**, 1049 (2003).

57. S. Ahmad, *The Self-Organizing Soot*, chapters 6-8, Amazon (2022).

58. A. N. Kolmogorov, *Information theory and the theory of algorithms. Selected works, Vol. 3* (Kluwer, Dordrecht, 1993), chapters 4 and 10.

59. S. Watanabe, *Knowing and Guessing*, John Wiley & Sons, New York (1969).

60. R. Y. Rubinstein and D. P. Kroese, "*The Cross-Entropy Method: A Unified Approach to Combinatorial Optimization, Monte-Carlo Simulation, and Machine Learning,*" Springer-Verlag, New York (2004).

61. A. Mao, M. Mohri, and Y. Zhong, "Cross-entropy loss functions: Theoretical analysis and applications," ICML 2023. https://arxiv.org/pdf/2304.07288.pdf

62. de Boer, D. P. Kroese, S. Mannor and R. Y. Rubinstein, "A tutorial

on the cross-entropy method," Annals of Operations Research, **134**, 19 (2005).

63. T. M. Cover, and J. A. Thomas, *Elements of Information Theory*, John Wiley & Sons, Inc, New York (1991).

64. S. Kullback, "Information Theory and Statistics," John Wiley & Sons., New York (1959).

65. J. R. Pierce, "An Introduction to Information Theory: Symbols, Signals and Noise," 2nd Ed. Dover Publications, Inc., New York (1980).

66. D. J. C. MacKay, *Information Theory, Inference and Learning Algorithms*," Cambridge University Press, Cambridge (2003).

67. Amari, Shun-ichi. "Information Geometry and Its Applications," Applied Mathematical Sciences. **194**, pp. XIII, 374. Springer Japan (2016).

68. I. J. Good, "Maximum entropy for hypothesis formulation, especially for multidimensional contingency tables," Ann. of Math. Statistics, 1963

69. S. Kullback and R. A. Leibler, "On information and sufficiency". Annals of Mathematical Statistics, **22**, 79 (1951).

70. The World Health Organization (WHO), "Summary of probable SARS cases by onset of illness from 1 November 2002 to 31 July 2003". https://www.who.int/publications/m/item/summary-of-probable-sars-cases-with-onset-of-illness-from-1-november-2002-to-31-july-2003

71. The WHO 2003 Case definitions for surveillance of severe acute respiratory syndrome (SARS). Daily situation Reports of cases and deaths. http://www.who.int/csr/sars/casedefinition/en/

72. J. D. Cherry and P. Krogstad, "SARS: The First Pandemic of the 21st Century", Pediatric Res. **56**, No. 1 (2004).

73. S. H. Lee, "The SARS epidemic in Hong Kong--a human calamity in the 21st century", Methods Inf. Med. **44**, 293 (2005).
74. J. S. Peiris and Y. Guan, "Confronting SARS: a view from Hong Kong," Philos. Trans. R. Soc. Lond. B Biol Sci. **359**, 1075 (2004).
75. Department of Health Hong Kong, "Outbreak of severe acute respiratory syndrome (SARS) at Amoy Gardens, Kowloon Bay, Hong Kong, main findings of the investigation," (2003). http://www.info.gov.hk/info/ap/pdf
76. Centers for Disease Control and Prevention, "Severe acute respiratory syndrome—Singapore," MMWR Morb. Mortal Weekly Rep **52**, 405 (2003).
77. Centers for Disease Control and Prevention, "Cluster of severe acute respiratory syndrome cases among protected health-care workers—Toronto, Canada," MMWR Morb. Mortal Weekly Rep **52**, 433 (2003).
78. S. Cauchemez et.al., "Real-time estimates in early detection of SARS," Emerg. Infect. Dis. **12**, 110 (2006).
79. E. Lau, "A comparative epidemiologic analysis of SARS in Hong Kong, Beijing and Taiwan," BMC Infectious Diseases, **10**, 50 (2010).
80. The complete lists of the COVID-19 cases are available on the WHO web page https://data.who.int/dashboards/covid19/.
81. The World Health Organization (WHO), Novel Coronavirus (2019-nCoV), Situation Report -1 and 2, 21st and 22nd January 2020, and the later ones issued daily.
82. COVID-19 in Korea: https://en.wikipedia.org/w/index.php?title=COVID-19_pandemic_in_South_Korea&oldid=962801750
83. COVID-19 in Germany:

https://en.wikipedia.org/w/index.php?title=COVID-19_pandemic_in_Germany&oldid=962549563

84. COVID-19 in Spain: https://en.wikipedia.org/w/index.php?title=COVID-19_pandemic_in_Spain&oldid=962697816

85. Our World In Data is a project of the Global Change Data Lab, a registered charity in England and Wales; https://ourworldindata.org/grapher/covid-variants-bar [our world].

86. https://ourworldindata.org/grapher/covid-variants-bar?time=2021-03-01&country=USA~DEU~AUS~KOR~CHN

87. P. V. Markov, et.al. "The evolution of SARS-CoV-2", Nature Reviews Microbiology, **21**, 361, (2023).

88. A. M. Carabelli, et. al. "SARS-C-V-2 variant biology: immune escape, transmission and fitness", Nature Reviews Microbiology, **21**, 162 (2023).

89. J. Y. Choi and D. M. Smith, "SARS-CoV-2 Variants of Concern", Yonsei Med J, **62**, 961 (2021).

90. A. Bolze, A. et al. "SARS-CoV-2 variant Delta rapidly displaced variant Alpha in the United States and led to higher viral loads", Cell Rep. Med. **3**, 100564 (2022).

91. W. T. Harvey, et al. "SARS-CoV-2 variants, spike mutations and immune escape", Nat. Rev. Microbiol. **19**, 409 (2021)

92. S. Pei, et.al. "Burden and characteristics of COVID-19 in the United States during 2020", Nature, **598**, 334 (2021).

93. K. Tao, et al. "The biological and clinical significance of emerging SARS-CoV-2 variants", Nat. Rev. Genet. **22**, 757 (2021).

94. H. Kalish, et al. "Undiagnosed SARS-CoV-2 seropositivity during the first six months of the COVID-19 pandemic in the United States", Sci. Transl. Med. **13**, 3826 (2021).

95. C. Y. S. Lee and J. B. Suzuki, "'COVID-19: Variants, Immunity, and Therapeutics for Non-Hospitalized Patients", Biomedicines **11**, 2055, (2023).
96. L. G. Thorne, et al. "Evolution of enhanced innate immune evasion by SARS-CoV-2", Nature **602**, 487 (2022).
97. S. Hakki, et al. "Onset and window of SARS-CoV-2 infectiousness and temporal correlation with symptom onset: a prospective, longitudinal, community cohort study", Lancet Respir. Med. **10**, 1061 (2022).
98. J. T. McCrone, et al. "Context-specific emergence and growth of the SARS-CoV-2 Delta variant", Nature **610**, 154 (2022).
99. M. Ghafari, et.al. "Investigating the evolutionary origins of the first three SARS-CoV-2 variants of concern", Front. Virol. **2**, 942555 (2022).
100. L. Subissi, et al. "An early warning system for emerging SARS-CoV-2 variants", Nat. Med. **28**, 1110 (2022).
101. Obermeyer, F. et al." Analysis of 6.4 million SARS-CoV-2 genomes identifies mutations associated with fitness", Science 376, 1327–1332 (2022).
102. S. Mallapaty, "Where did Omicron come from? Three key theories," Nature **602**, 26 (2022).
103. J. J. Dennehy, et.al. "Where is the next SARS-CoV-2 variant of concern?", Lancet **399**, 1938 (2022).
104. F. Ndaïrou, "Mathematical modeling of covid-19 transmission dynamics with a case study of Wuhan", Chaos Solitons Fractals **135**, 109846 (2020).
105. A. R. Fehr, and S. Perlman, "Coronaviruses: an overview of their replication and pathogenesis," Methods Mol. Biol. **1282**, 1 (2015).

106. J. Cui, F. Li, and Z. L. Shi, "Origin and evolution of pathogenic coronaviruses," Nat. Rev. Microbiol. **17**, 181 (2019).
107. Y. Zhang, S. Tang and G. Yu, "An interpretable hybrid predictive model of COVID-19 cases using autoregressive model and LSTM", Scientific Reports, **13**, 6708 (2023).
108. S. Khoojine, et.al. "Network autoregressive model for the prediction of covid-19 considering the disease interaction in neighboring countries", Entropy **23**, 1267 (2021).
109. M. Yadav, M. Perumal and M. Srinivas, "Analysis on novel coronavirus (covid-19) using machine learning methods", Chaos Solitons Fractals 139, 110050. (2020).
110. A. Antoniadis, S. L. Lacroix, and J. M. Poggi, "Random forests for global sensitivity analysis: A selective review," Reliability Engineering & System Safety, **206**, 107312 (2021).
111. C. M. Yeşilkanat, "Spatio-temporal estimation of the daily cases of covid-19 worldwide using random forest machine learning algorithm", Chaos Solitons Fractals 140, 110210 (2020).
112. D. Chumachenko, et.al. "Investigation of statistical machine learning models for covid-19 epidemic process simulation: Random forest, k-nearest neighbors, gradient boosting", Computation **10**, 86 (2022).
113. J. Galasso, D. M. Cao and R. Hochberg, "A random forest model for forecasting regional covid-19 cases utilizing reproduction number estimates and demographic data", Chaos Solitons Fractals 156, 111779 (2022).
114. C. M. Yeşilkanat, "Spatio-temporal estimation of the daily cases of covid-19 worldwide using random forest machine learning algorithm", Chaos Solitons Fractals **140**, 110210 (2020).
115. J. Shaman, and M. Galanti, "Will SARS-CoV-2 become

endemic?" Science **370**, 527 (2020).

116.　A. Varghese, et.al. "A global perspective on the intrinsic dimensionality of COVID-19 data", Scientific Reports, **13**, 9761 (2023).

ABOUT THE AUTHOR

Shoaib Ahmad earned his PhD in physics from Sussex University in UK. He worked and taught physics at British and Pakistani universities and research institutions. He held the Professor Rafi Chair of Physics at GC University, Lahore. Besides publishing in physics journals, he writes on science, energy, and technological issues for the non-specialists, in Urdu and English. In recent years, he has been working on self-organizing, information generating and manipulating dynamical systems and developing information-theoretic models for the emerging dynamical systems. His previous book *'The Self-Organizing Soot'* published on Amazon in 2022, investigates the emergence of the nanostructures of Carbon. In the present book, the information-theoretic 'Source-Reservoir-Sink' model, based on the probability distributions of the emerging, dissipative structures and events, generates the comprehensive profiles of the information-generating pandemics. Here, in this book on the *'Cross Emergence of COVID-19'*, the information-theoretic diagnostic tools developed for COVID-19's dynamical system in the form of Emergence and Cross Emergence Functions and the Indices of Divergence, quantify the Evolution of the self-organizing SARS-CoV-2.

www.ingramcontent.com/pod-product-compliance
Lightning Source LLC
Chambersburg PA
CBHW052149220526
45471CB00004B/1595